Ballon Hoch

The German Ballontruppe in WWI

Michael Schmeelke

Translated by Catherine Lawler

Great War Aviation Centennial Series #91

Publisher Notes
For our aviation books, please see our website at: **www.aeronautbooks.com**.
I am looking for photographs of the less well-known German aircraft of WWI. For questions or to help with photographs you may contact me at **jherris@me.com**

ISBN: 978-1-953201-85-0

Profiles: Bob Pearson, David Méchin, Alexsandr Kazakov
Design and layout: Jack Herris
Cover design: Aaron Weaver
Digital photo editing: Jack Herris

Aeronaut Books
Books for Enthusiasts by Enthusiasts
www.aeronautbooks.com

Table of Contents

Preface

Above: Behind a destroyed Fokker E. III, the crew is seen preparing a balloon ascent.

When brothers Jacques-Étienne and Joseph-Michel Montgolfier test flew their first balloon in France on June 4, 1873, they had created the prototype for all „lighter than air" category balloons. A variety of successful balloon rises by Pilâtre de Rozier, the Prince of Ligne, and the Dukes of Laurencin followed suit. Rozier and Romain were among the first ballooning victims. They fatally crashed near Wimereux on June 15, 1874 when their hydrogen balloon caught fire.

In the US, John Wise had completed a long range balloon trip covering a distance of 1283 kilometers in 1859; he planned an Atlantic crossing at a later time which did not come to pass due to the outbreak of the American Civil War. Wise also invented the rip panel (Reissbahn), a device that would tear open the balloon casing allowing hydrogen to escape in the event of an emergency. This invention formed part of the standard balloon equipment until the 1940's.

Aerospace visionary Count Zeppelin mentioned a balloon vessel in his diary in 1874; contrary to previous models, however, Zeppelin's reference was to a steerable balloon. He kept track of the balloon missions of the nothern states during the Civil War and completed his first balloon ascent near St. Paul, Minnesota on August 18, 1863. During the Franco-Prussian War of 1870-71, Zeppelin witnessed the escape of government officials from occupied Paris using balloons, an act that was not preventable by the Prussian troops. The French Army used such balloons for reconnaissance.[1]

Following the experiences of the Franco-Prussian War of 1870-71, the German Reich began developing its own balloon troop. On May 9, 1894 the Prussian Army established the *Ballon-Détachement,* an experimental station for *Captiv-Ballons* led Hptm. Buchholz. Mid 1890, the airship department was separated from the Eisenbahntruppe and Luftschiffer-Bataillon 1 was established in 1901,and a Bavarian Bataillon in 1904; both were based on the 1899 *Präsenzgesetz.*

A total of 302 soldiers formed the Fesselballontruppe during peace time.[2] From 1896, Army balloons (Kugelballone) were replaced by Parseval-Sigsfeld balloons (Feldballone). The new balloons contained 600-800 cubic meters of hydrogen and served as a reconnaissance platform for observation and photographic documentation of enemy territories. A telephone apparatus was used for direct contact

1 Schiller, Hans von: Zeppelin – Wegbereiter der Luftfahrt, Bad Godesberg 1966.

2 National Archives: Kriegsrüstung und Wirtschaft, Berlin 1930.

Above: Impressed spectators observing the ascent of a balloon in France.

to ground personnel. Balloons were routinely utilized during exercises of the Imperial Army as well as at its troop exercise location at Döberitz[3] where mobile warfare was frequently practiced. As a result, there was hardly any collaboration between General staff and artillery.

Due to the rapid advancements in aircraft development during that time, expansion and technical progress of balloon troops was neglected until the beginning of the war in August of 1914. Examinations of hydrogen permeability in rubberized balloon materials and the invention and development of overpressure valves were primarily utilized for the construction of airships but balloon troops also benefited.[4]

It is not surprising that the German General Staff did not issue orders for Feldluftschiffer units to mobilize to towards the Western and Eastern Fronts in July 1914. Balloons were not expected to contribute to the successes of the expected short mobile warfare effort. General-Major Wilhelm Messing, Chef der Inspektion des Militär- Luft- und Kraftfahrwesens who oversaw the balloon troop, dissented. Eventually,

Chef des Generalstabes, Generaloberst von Moltke, ordered the mobilization of eight Feldluftschiffer units and 15 Festungs-Luftschiffer troops. A horse-drawn Feldluftschiffer unit was assigned to each Army department. It usually consisted of one balloon, at times two. In October 1914, the German Army was reorganized in the West. According to an Armed Forces list dated October 10, 1914 each Army branch was mentioned along with its assigned

3 Biermann, Kai & Cielewicz: Flugplatz Döberitz, Berlin 2005.

4 Frenzel, Walter: Über die Gasdurchlässigkeit der gummierten Ballonstoffe, Elsässisches Textilblatt, Gebweiler 1912.

troops; balloon units however were not listed. This indicates how little interest the OHL showed in Feldluftschiffer units at the time. The following five years of war however would show the value of balloons to the military in spite of rapididly evolving aircraft developments, especially when mobile warfare turned into trench warfare.

By the end of the war, the German Army included 184 balloon troops. In addition to the Bavarian departments, the Land of Württemberg founded the Feldluftschiffer-Abteilung (FLA) which consisted of FLA 30 with balloon troops (Ballonzüge or Bz) 78, 79, and 94 as well as the Saxon FLA 5 which consisted of Bz 20 and 32. As an ally to Germany, Austria-Hungary established 32 balloon troops before the end of the war.

Balloon troop literature is usually limited to short accounts written by individual balloon observers with the exception of a book written by Lt. d.R. Ernst Struck who published his experiences as a balloon observer during the war.[5]

In 1986, I received a cardboard box filled with balloon photosgraphs taken by former balloon observer Richard Wagenscheid. Unfortunately, his nephew, Dr. Eberhard Ulbert was only able to supply very basic information regarding his uncles' military career. A handful of written documents along with the photographs collected indicate that Lt. Wagenscheid served with Bavarian Feldluftschiffer-Abteilung 8/208. After NSDAP assumed power in 1933, Richard Wagenscheid initially seemed enthusiastic about the new government. Having grown up during a monarchy, he was slow to adjust to the liberal politics of the Weimar Republic. Like many Germans during that time, he did not agree with the Versailles peace treaty; soon, Wagenscheid discovered that this regime did not have much in common with his point of view. Ultimately, he publicly opposed to the Third Reich. As a result, an SA troop entered his apartment and murdered him.

I was able to supplement Richard Wagenscheid's extensive inheritance using personal photographs taken by former balloon observers Lt. Thofehrn and Lt. Kerstenhann. Unfortunately, any written documents had been destroyed. The images show that Lt. Thofehrn trained with the 1st Luftschiffer-Bataillon in Jüterbog in October of 1915; he was later stationed in France's Champagne region. The inheritance also contained aerial images from the home area which appear to have been taken during training exercises. It is possible that Lt. Thofehrn was stationed in the Bosporus region as indicated by some images which show the Hagia Sophia, a Byzantine church, as well as Osman warships. Furthermore, I used images found in photo albums owned by

5 Struck, Ernst: Im Fesselballon, Berlin 1918.

Above: Memorial stone the cemetery in Wimille in remembrance of Pilátre de Rozier and Pierre-Ange Romain, who fatally crashed on June 15, 1874.

Surgeon General d. L Prof. Dr. V. Hofmeister who served on the front of the Somme in 1916.

Writing this book would not have been possible without the help and support of Dr. Eberhard Ulbert, Greg VanWyngarden, and Bruno Schmäling. Their guidance and resources, particularly the many references and rare images were instrumental to my completing the book. I would also like to express my gratitude to the US National Archives and Records Administration (NARA) for supplying many rare documents and photographs. Our publisher, Jack Herris has again facilitated making another chapter of aviation history accessible to the public. Thank you, Jack. Catherine Lawler translated this book which must not have been an easy task given the many military terms. My sincere gratitude goes to all of you.

Michael Schmeelke

ETABLISSEMENTS

Above: Ballast bags attached to the balloon basket.

Facing Page: Filling hole for a Kugelballon.

Albatros D.III(OEF) 153.10
***Hptm*. Godwin Brumowski**
OC, *Flik* 41/J
35 total victories, 5 balloons

The Balloon Squadron in the First Year of the War

Above: Luftschiffer training at Berlin Tegel (Luftschiffer-Bataillon 1) using Kugel balloons.

As previously mentioned, the Luftschiffer Squadron was dormant until the beginning of the war. In fact, the dissolution of the squadrion was anticipated both in Germany and in France. The General-Inspektion des Militär-Verkehrswesens which was the highest Central Authority showed little interest in the Fesselballontruppe. The subordinate authority, Inspektion des Militär- Luft- und Kraftfahrwesen (led by General-Major Wilhelm Messing) was subdivided into Inspektion der Luftschiffer-Truppe (Iluft) commanded by Oberst Kuno v. Barfus and the Fliegertruppe, barely received any funding for the Ballontruppe.[1] During the extended period of peace there was no need for enhancement of technical equipment since the establishment of a patent for a cylindrical balloon with a funnel shaped vestibule on October 10, 1883 by engineer August von Parseval and Hptm. d.R. Hans Bartsch von Sigsfeld.

The *Drachenballon* was introduced in 1896, thereby replacing the original round balloon. On February 1, 1912, Hptm. d.R. Bartsch von Sigsfeld fatally crashed in a free balloon (Freiballon) in Antwerp.

During the Prussian armed forces reinforcement efforts in 1913, Luftschiffer Battalions were increased from two to five. Out of the five Prussian Battalions listed in the Royal Prussian Army's rankings, only Luftschiffer Battalion 1 was responsible for training of balloon personnel. The other four battalions were mainly responsible for airships which continued to be privileged by the general and naval staff. The Bavarian airship battalion was subordinate to the Luft- and Kraftfahr-Bataillon (aviation and transportation battalions) in Munich, led by Major Karl Reitmeyer.[2]

When the troops were mobilized from August 2, 1914 German Fesseballon squadrons were renamed Feldluftschiffer squadrons. The tactical unit name was Feldluftschiffer-Abteilung (FLA). Seven Prussian and two Bavarian FLA[3] moved to the Western and Eastern front as planned from the 7th through the 9th days of mobilization. Each squadron consisted of two Zügen which had one Drachenballon and one Reserveballon each.[4] Individual Züge were also known as Halbabteilung (half division). The gas convoy (Gaskolonne) had an additional reserve balloon. As there was only one winch available, only one balloon per Zug was able to rise at any given time. Each FLA consisted of 10 officers, 270 soldiers, 194 horses, and 35 vehicles.

Above: Ascent of a Bavarian Luftschiffer-Bataillon balloon near Munich.

Aufstellung der Festungs - Luftschiffer-Truppe 2 August 1914		
Created From	**Number**	**CO**
Kommandantur Köln	Festungs-Luftschiffertrupp 10	Hptm. Mickel
Kommandantur Neu-Breisach	Festungs Luftschiffertrupp 13	Oblt. Bachmayr
Pionier Btl. 15 Straßburg	Festungs-Luftschiffertrupp 14	Hptm. Batzer
Pionier Btl. 15 Straßburg	Festungs-Luftschiffertrupp 15	Rittmeister d.R. Bartmann
Kommandantur Metz	Festungs-Luftschiffertrupp 18	Oblt. Rudersdorf
Kommandantur Metz	Festungs-Luftschiffertrupp 19	Oblt. Neidhardt
Kommandantur Metz	Festungs-Luftschiffertrupp 20	Oblt. Wolffenstetter
Kommandantur Metz	Festungs-Luftschiffertrupp 21	Oblt. Schmidt
Kommandantur Diedenhofen	Festungs-Luftschiffertrupp 22	Oblt. Moller
Pionier Btl. 25 Mainz	Festungs-Luftschiffertrupp 29	Oblt. Möschke
Gouvernement Königsberg	Festungs-Luftschiffertrupp 1	Oblt. v. Kyckbusch
Gouvernement Posen	Festungs-Luftschiffertrupp 5	Hptm. Meyer
Gouvernement Thorn	Festungs-Luftschiffertrupp 23	Hptm. Granier
Gouvernement Thorn	Festungs-Luftschiffertrupp 24	Oblt. d.L. Link
Gouvernement Graudenz	Festungs-Luftschiffertrupp 26	Hptm. V. Gellhorn
By. Btl. München	By. Festungs-Luftschiffertrupp 1 Germersheim	Hptm. Koban

Luftschiffer-Batallion 1 in Berlin-Tegel and the Bavarian Batallion in Schleißheim were designated replacement batallions which were responsible for the replenishment of personnel and materials. Initially, the 15 established Festungs-Luftschiffer squadrons were unsuitable for an assignment at the front as they did not have enough horses or vehicles.[5] They consisted of 5 officers, 53 horses, 8 vehicles, and 1 balloon.

Luftschiffer-Truppe personnel consisted of many volunteers, including some Freiballonfahrer with prior experience. From 1900 until 1914, balloon clubs were established all across Europe; their flights drew crowds of balloon enthusiasts and spectators. The German Reich promoted ballooning by allowing the transport of balloons by railway at a discounted military rate. More than 100 races were held annually across Germany and France. However, only wealthy individuals were able to afford to participate in this expensive and elitist sport.

In 1914, there were 694 balloon pilots in Germany, many of which were drafted to the Luftschiffertruppe when the war broke out. Pilots of the Ballon-Sport-Verein Leipzig signed up to serve with the Luftschiffertruppe immediately following the mobilization:[6]

"Wilhelm Fahlbusch, Alfred Restler, Fritz Bertram, Hubert Schreiterer, and Gerhard *Kühne are all from Leipzig and are now serving with our balloon squadron.*"

In France, the birthplace of balloons, dissolving the Compagnie d'Aérostiers continued to be a consideratin in early 1914. However, the balloon-captifs of Albert Irénée Caquot (type P, with a volume of 750 m^3) soon turned into an indispensable reconnaissance equipment within the French army. Caquot continued to develop the ballons, evolving to type H in 1915 (800 m^3) and type L (880 m^3). Personnel were trained at Saint-Pol, Notre-Dame-de-Lorette. From May 1, 1916, training was also completed at the balloon training facility in Vadenay, near Châlons. Thanks Delahay and Latil trucks which were outfitted with motor winches, the Compagnie d'Aérostiers could be deployed with high degree of flexibility[7] while Austrian-Hungarian Luftschiffer continued to deploy with horses and manual winches.

In May of 1915, the French Army Command assigned 40th Compagnie d'Aérostiers to serve as the first ballooning unit with the British Army. Later, the RFC/RAF, and RNAS balloon wings were formed from the French unit. They utilized the Caquot balloon types which was also produced in England by CG Spencer & Sons, Vickers Ltd, and Airship Ltd. In 1916, 800 m^3 and 1000T m^3 balloons[8] were replaced

Prussian Luftschiffer-Bataillone in May 1914

Bataillon	Town	Commander
Luftschiffer-Bataillon Nr. 1	IP	Major Krenzlin
	Berlin Reinickendorf	
Luftschiffer-Bataillon Nr. 2	Berlin-Tegel	Major Schulze
Stab + 1. Kompanie	Berlin-Tegel	
2. Kompanie	Königsberg	
3. Kompanie (sächsisch)	Dresden	
Luftschiffer-Bataillon Nr. 3	Köln	Major v. Schulz
Stab + 1. Kompanie	Köln	
2. Kompanie	Düsseldorf	
3. Kompanie	Metz	
Luftschiffer-Bataillon Nr. 4	Mannheim	Major v. Bockelmann
Stab + 1. Kompanie	Mannheim	
2. Kompanie	Metz	
3. Kompanie	Gotha	
4. Kompanie (württ)	Friedrichshafen	
Luftschiffer-Bataillon Nr. 5	Graudenz	Major Knappe
Stab + 2. Kompanie	Liegnitz	
1. Kompanie	Allenstein	
3. Kompanie	Posen	

Above: Drachen balloon equipped with dual Kegel steering, built by Fa. August Riedinger, Munich, 1897.

Aufstellung der FLA - Luftschiffer-Truppe 2 August 1914				
Created From	**Location**	**FLA No.**	**Armee**	**CO**
Luftschiffer-Bataillon 1	Berlin	FLA 1	1. Armee	Hptm. von Zychlinski
Luftschiffer-Bataillon 3	Köln	FLA 2	2. Armee	Hptm. Spangenberg
Pionier-Bataillon 21	Mainz-Kastell	FLA 3	4. Armee	Hptm. Schoof
Infanterie-Regiment 46	Posen	FLA 4	5. Armee	Hptm. Stottmeister
Pionier-Bataillon 3	Spandau	FLA 5	?	Hptm. Platzhoff
Pionier-Bataillon 15	Straßburg	FLA 6	7. Armee	Hptm. Kalsow
Infanterie-Regiment 28	Koblenz	FLA 7	3. Armee	Hptm. Mezel
Luftschiffer-Bataillon 5	Königsberg	FLA 8	8. Armee	Hptm. Schellbach
Bay. Luft- und Kraftfahr Btl.	München	Bay. FLA	6. Armee	Hptm. Lochmüller

7. Luftschiffertruppen.

	Stichjahre	Kopfstärke: Offz. (einschl. San.- u. Veter.-Offz.)	Kopfstärke: Unteroffz. (einschl. Unterbeamte)	Kopfstärke: Mann	Kopfstärke: San.- u. Trainpersonal	Waffen: Gewehre bzw. Karabiner	Waffen: Revolver bzw. Pistolen	Fesselballons	Fahrzeuge: Gaswagen	Fahrzeuge: Gerätewagen	Fahrzeuge: Windewagen	Fahrzeuge: Fernsprechwagen	Fahrzeuge: Kraftwagen	Große Bagage: Packwagen	Große Bagage: Lebensmittelwagen	Große Bagage: Futterwagen	Pferde: Reit-	Pferde: Zug-
(Feld-) Luftschiffer-Abt.	1888	2	9	53	17	60	1	2	2	2	1	—	—	—	—	—	2	26
	1898 bis 1902	9	11	96	70	103	58	2	12	2	1	—	—	1	1	1	20	102
	1910 bis 1914	8	11	96	70	103	60	2	12	2	1	1	-	1	1	1	19	104
Festungs-Luftschiffer-Abt.	1902	5	6 oder a) 7	65 oder a) 77	5 oder a) 9	71 oder a) 84	—	1	0 oder a) 6	1	1	—	—	—	—	—	5	8 oder a) 16
	1910	5	6 oder a) 7	65 oder a) 77	5 oder a) 25	71 oder a) 84	—	1	0 oder a) 6	1	1	—	—	—	—	—	5	8 oder a) 48
	1914	5	7	77	25	84	—	1	6	1	—	—	—	—	—	—	5	48
Feldtrupp f. Lenkluftsch.	1910	3	7	75	3	82	—	—	—	—	—	—	—	1	—	—	2	2
Luftschifftrupp . . .	1914	3	10	131	5	139	1	—	—	—	—	—	3 b)	1	—	—	3	2
Ersatz-Abt. mit Rekr. Dep. . .	1888	4	9	52	1	58	—	—	—	—	—	—	—	—	—	—	3	—
	1898	9	20	204	1	222	—	—	—	—	—	—	—	—	—	—	3	1
„ „ „ und Besp. Abt. . . .	1902 bis 1910	15	41	420	2	415	37	—	6	1	1	—	—	—	—	—	19	44
mit Rekr. Dep. . .	1914	6	23	195	1	212	—	—	—	—	—	—	—	—	—	—	1	—
Ers. Besp. Abt. . . .	1914	1	7	30	1	—	37	—	6	1	1	—	—	—	—	—	14	44

a) Kugel- oder Drachenballon. — b) 1 Personen-, 2 Lastkraftwagen.

Above: Luftschiffer-Truppe development until 1914.

by Caquot Type M vessels which had been introduced to Compagnie d'Aérostiers in 1915. This streamlined balloon contained three stearing bladders, rising to an altitude of 1500 m with two observers.

The American Expeditionary Forces (AEF) balloon troop arrived at Le Havre by ship on December 28, 1917. The soldiers had been well trained at the ballon training facility in Ft. Omaha, Nebraska. Later, front training was carried out by experienced members of the Compagnie d'Aérostiers in Cuperly near Vadenay. Thereafter, personnel were temporarily assigned to serve on the front with the French Compagnie d'Aérostiers in order to gain practical experience. AEF

Facing Page: Bavarian Luftschiffer-Bataillon 1 uniform coat.

Above: Oberst Kuno v. Barfus was in command of Inspektion der Luftschiffer-Truppe (Iluft).

Left: General-Major Wilhelm Messing, 1914 Chief of Inspektion des Militär- Luft-, and Kraftfahrwesens.

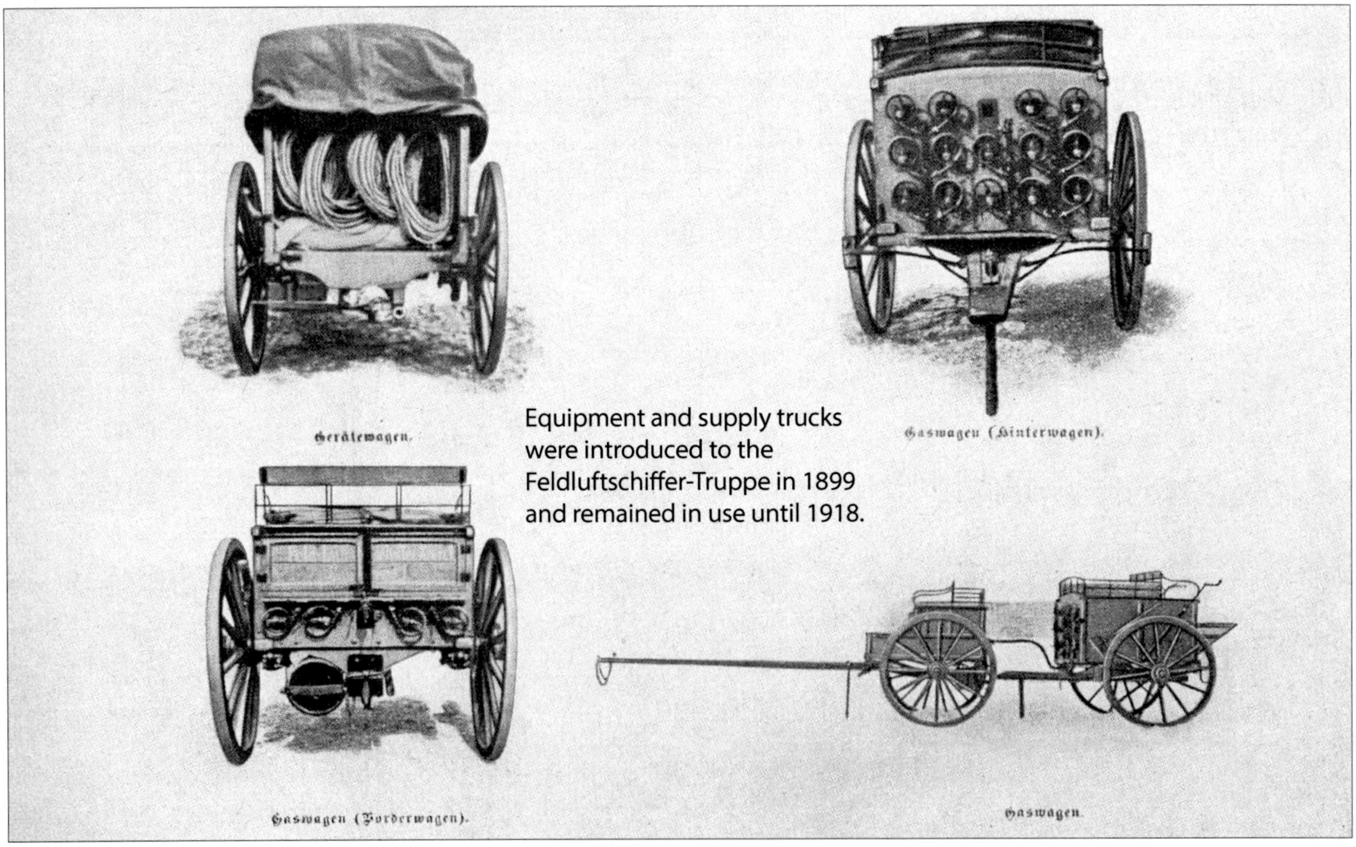

Equipment and supply trucks were introduced to the Feldluftschiffer-Truppe in 1899 and remained in use until 1918.

Above: Kugel balloons were also used to drop propaganda materials, such as newspapers in French or English, over the enemy Hinterland.

Above: After the mobilization, the Bavarian FLA 8 left Munich to go to war.

Above: A Parseval-Sigsfeld Ballon is ready for ascent. After the balloon was filled with gas, the horse-drawn gas carriages were moved to a safe distance.

utilized its balloon troops more extensively for the first time in the battle of Belleau-Wood. French Generals remained in Command as the French did not yet entrust American Generals with offensive leadership. American Generals were not put in charge of overall troop command until the "all American battles" near St. Mihiel and Meuse-Argonne.[9]

The Feldluftschiffer-Truppe on the Western Front

Until mid August 1914, mobile Feldluftschiffer-Abteilungen (FLA) only rarely appeared with the Army on the Front. They struggled to keep up with German troops rushing forward using their 600 - 800 m^3 tethered balloons as their equipment was at the level of 1901.[10] Furthermore, horses were in short supply during the mobilization which led to the delayed arrival of FLA 2 of the 2nd Army (Hptm. Spangenberg) to the capture of fort Lüttich. There, Feldflieger-Abteilungen 3, 9, and 11 took over reconnaissance missions for the time being. Feldschiffer-Abteilung did not noticeably support army troups with reconnaissance work until August 15, when FLA 2 was in command of artillery during the bombardment of the Western forts at Lüttich (Boncelles, Lantin, and Loncin). Lüttich fell to the Germans the following day. On August 20, the Bavarian Feldluftschiffer-Abteilung (Hptm. Lochmüller) assigned to the 6th Army accurately aimed a

Left: A Kugel balloon ascends during a contest in France in 1912.

Above: A French winch wagon at the 5th Exposition Internationale de Locomotion Aérienne in Paris in December of 1913. This motor-driven winch enabled a quick ascent and collection of the balloons. The captions previously mentioned that balloons would no longer be utilized at that point. (Zeitschrift für Flugtechnik)

Above: A French, 750 m^3 Ballone-captif prior to ascent. Albert Irénée Caquot continued to develop these balloons during the war for the utilization by Allied Forces.

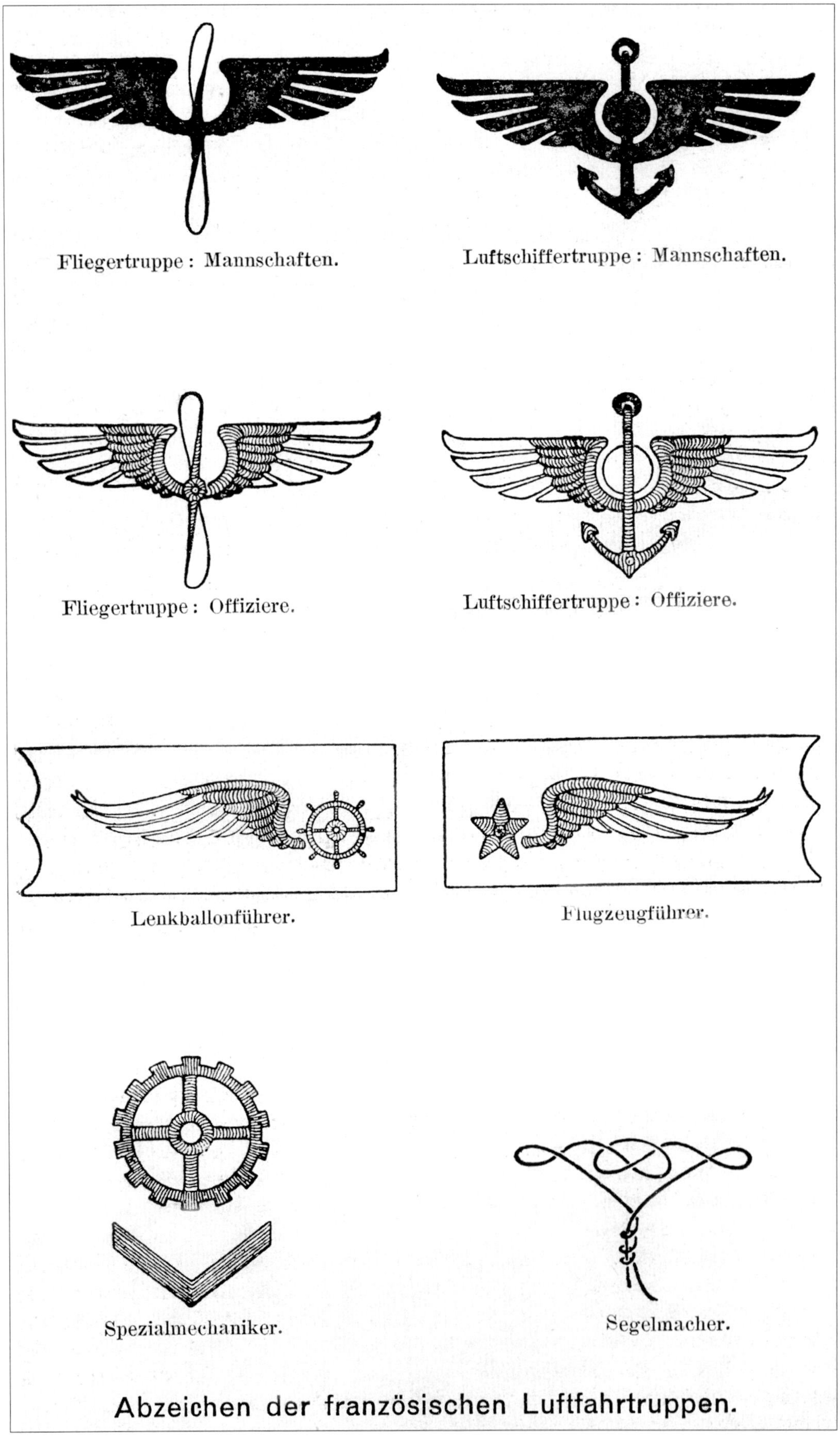

Left: Badge of the French Compagnie d'Aérostiers.

Motorized winch, mounted to a Delahaye truck.

Above: Motorized winch, mounted to a Delahaye truck.

Left: The ground crew moves a French, 750 m^3 Ballone-captif (Type P) toward its ascent point.

Right: Marcel Hutteau, observer of the 34th Compagnie d'Aérostiers in August 1916.

21 cm mortar battery at a French artillery position. Between 13:30 and 14:30, the battery fired multiple salvos assisted by the tethered balloon, until French guns went silent. On the following day, eith enemy cannons were discovered in the abandoned area. An observer stationed in the village of Kirberg reported the continued withdrawal of French troops extending back beyond the town of Saarburg.

On August 7, the 7th German Army was met by the French 1ème Armee who advanced to the South of the Western front from Belfort to Mühlhausen. By August 9, the 7th Army under the command of Generaloberst v. Heeringen succeeded in beating the French, forcing them to retreat to Belfort. Thereupon, the French Armee d'Alsace was staffed with fresh troops, amongst them the dreaded Chasseurs Alpines who advanced back to the Rhine river but ultimately withdrew again by August 25. The threat of a

Collection "Patrie"
G. DE GRAFFIGNY
La Saucisse
Infernale
40c.
Le récit complet
illustré.

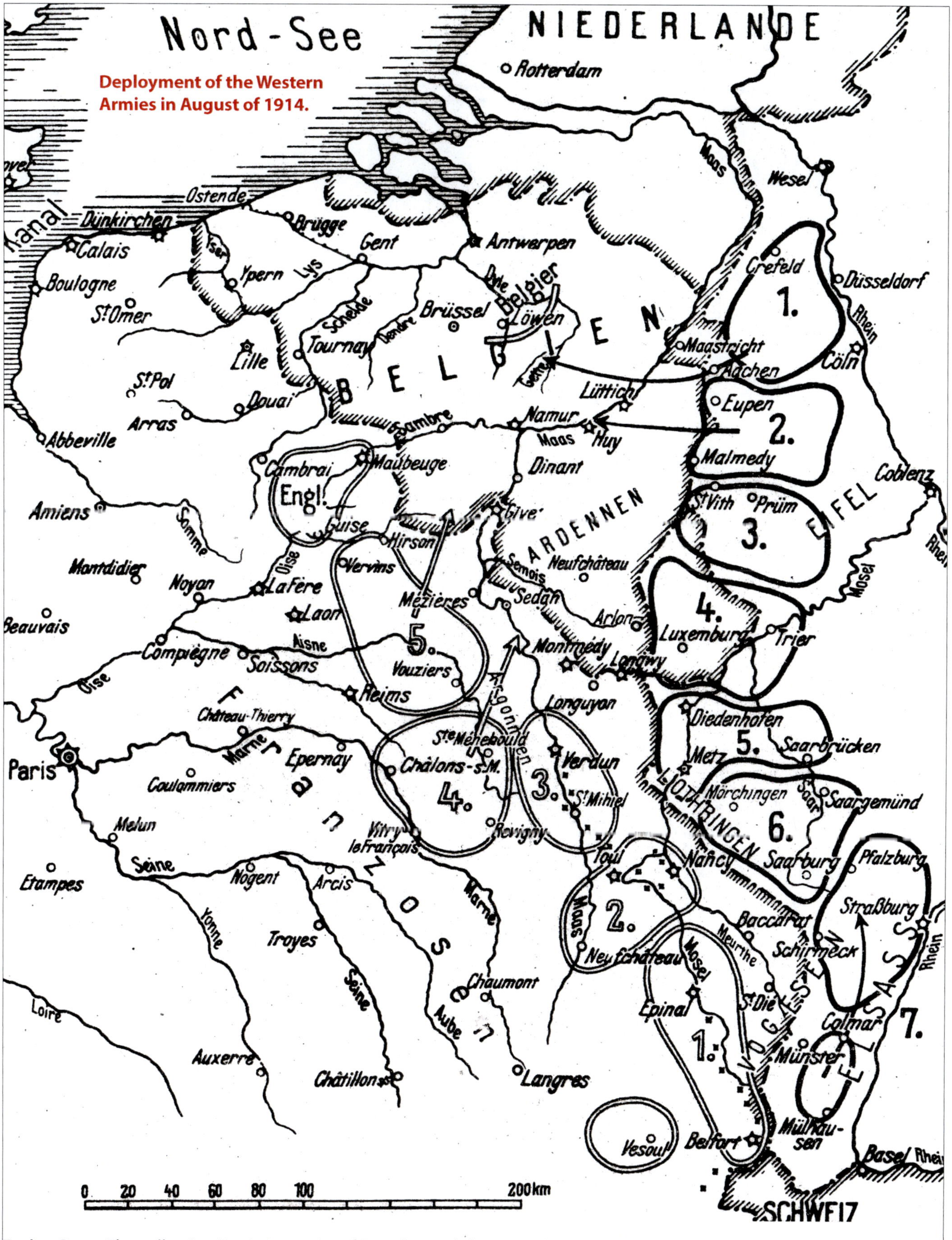

Deployment of the Western Armies in August of 1914.

Facing Page: The collection Patrie is a series of French novels honoring battles and soldiers during World War I. For a cost of 40 centimes, readers were able to learn everything there was to know about the work of the Compagnies d'Aérostiers.

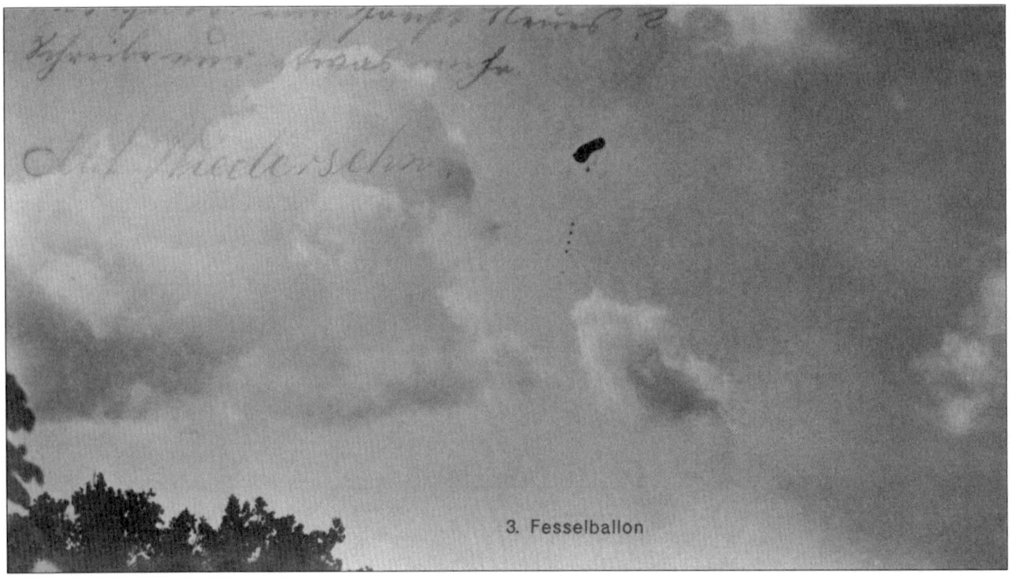

Left: German balloon on the Western Front, September 1914.

Facing Page, Above: Fort Champ des Romains after its capture by German troops.

Right: The American Expeditionary Forces (AEF) Ballontruppe underwent further training at Cuperly. The image shows a US Balloon Service observer prior to his ascent (NARA).

Facing Page, Below: Fort Camp de Romains was hard to hit with the K.u.K. artillery's 30.5 cm Skoda mortars.

Left: Aerial image of the Manonvillers stone block outer fort, August 22, 1914.

Grabenwehr Fort C. d. R.

VENNE IMPIEGAT
BZURA SUCKA
ZARKI - VHR - G
WATRA - JAKOB
CEARDARI - GORL
DOBERDO' - FOLG

Above: Until September 1916, Gefreiter Wilhelm Schmidt was part of König Karl 1's Württembergisches Feldartillerie-Regiment number 13.

further French offensive in the Alsace could not be ruled out in 1914. In August of 1914, the German Fliegertruppe in the Vosges consisted solely of the 8th Bavarian Feldflieger-Abteilung. On August 20, Festungs-Luftschiffertrupp 13 (Oblt. Bachmayr) from Breisach joined the Fliegertruppe to conduct aerial surveillance of the Vosges as well as the Rhine valley. Festungs-Luftschiffertrupp 14 (Hptm. Batzer) monitored the areas near Dahlenheim, north of Molsheim, and the Breusch valley. The French opposition immediately began combatting the German tethered balloons using flechettes launched by Escadrilles BL3 and BL10. The Escadrilles' Bleriot XI-2 and Maurice-Farman MF 7 were stationed at Belfort-Chaux airfield which was only a few minutes away by air.[11] While only little attention was paid to the Alsace as a theater of war at that time, the battles for the Vosges mountains as well as Hartmannsweiler- and Lingenkopf continued until 1918.

On August 22, 1914 the 31st Infanterie-Division (AOK 6) conquered the city of Lunéville. South of the city, the outer fort of Manonvillers was located; this area was to be captured by means of heavy artillery. The fire of the 21cm howitzers and 13cm cannons was directed by a balloon operated by the Bavarian Luftschiffer-Kompagnie until the garrison of the fort surrendered on August 27.[12] Following is an account written by war correspondent Heinrich Binder of Berliner Tageblatt newspaper, dated September 6, 1914[13]:

"To the left of the road to Aincreville was a giant balloon of a Luftschifferabteilung; it resembled a large yellow banana. The entire machine is a marvel of modern ability and skill. The cars beside it each contained 20 tanks of hydrogen. Each container has a volume of 5-6 m^3. It takes approximately 15-16 minutes from the time of unpacking until the fully filled balloon ascends about 100 meters into the air. [...] During the bombardment of the French outer fort of Manovillers from the border station of German Avricourt, our Luftschiffer rendered excellent service by operating one of their balloons as the effectiveness of our hits could be carefully observed from above and reported back to us."

Fort Camp-des-Romains which was part of the chain of the Verdun outer forts, fell thanks to the support of the Feldluftschiffer. The Army division (AA-Strantz) which was tasked with the capture of the outer forts was formed on September 16, 1914 from the IIIrd Bavarian Armeekorps, the Prussian V. Korps, and the 33rd Reserve-Division. For this purpose, heavy artillery was utilized, including 21 cm field howitzers, 28 cm coastal mortars, and 30.5cm Austrian mortars. As ground reconnaissance was challenging to carry out in the densely forested area, balloon observers were necessary to relay target instructions for the bombardment of the fort. On September 25, the troops of the 11th Infanterie-Regiment succeeded in conquering the fort.[14]

Infantry and artillery alike were not fond of the close proximity of the balloons. They feared that the enemy would recognize their intent to attack, directing heavy artillery fire towards their positions. Below is a report by Dr. Oskar Bongard of Vossische Zeitung, dated September 26, 1914:[15]

"Near St. Maurice, Côte Lorraine. The crews of a Luftschifferabteilung were in the process of getting down a tethered balloon which had been raised for reconnaissance purposes. Enemy artillery forces began to develop an interest in it. [...] Following a loud, sharp bang, the first shrapnel balls smashed to the ground 30 meters from us. Many more followed in rapid succession for a while. Soon, the shrapnel no longer burst above us but rather flaired and hissed above our heads, scattering their balls in the direction from which the balloon had risen."

On August 30, 1914 the Bavarian FLA 1 had moved to a balloon release location southwest of Sionviller. 70 meters in front of their location, a Prussian 13cm and two 10cm

Above: Thanks to reconnaissance work by the Bavarian FLA 1, this French 7.5cm battery position location east of Signeulx was hit by German artillery.

Above: Aerospace pioneer Adolphe Pegoud departs for a bomb drop mission over German troops in Bleriot XI.

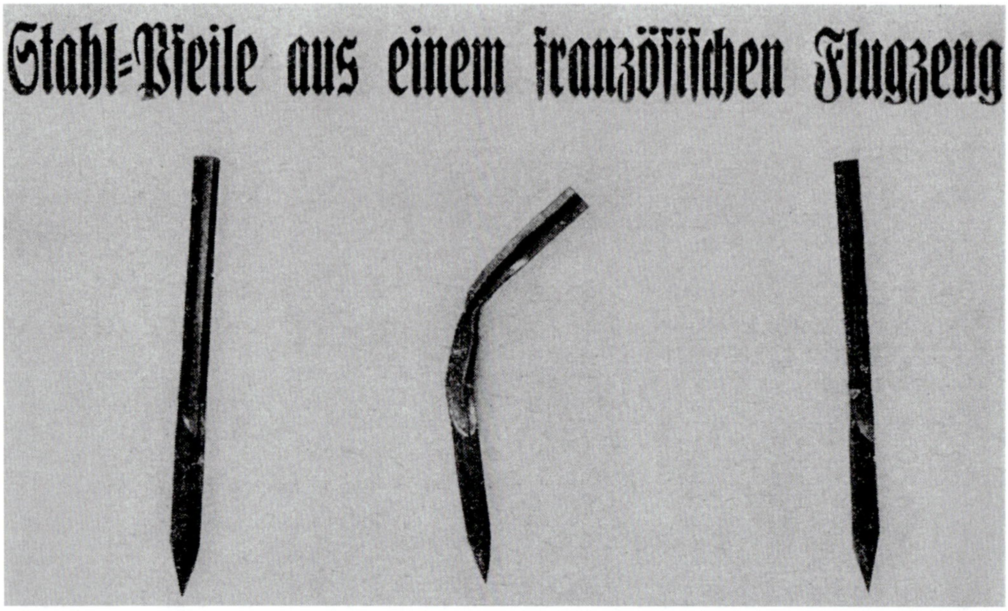

Left & Below: French pilots used flechettes to down the first German balloon on September 1, 1914.

Bottom: 21 cm mortars arriving in a field without air cover; during the first weeks of the war; the dangers from the French enemy were still unknown.

Above: In the fall, the German artillery was lacking ammunition. Therefore, German artillery soldiers frequently commandeered materials from defeated enemies, such as this 12 cm French cannon including ammunition. Artillerists are seen watching the ascent of their balloon.

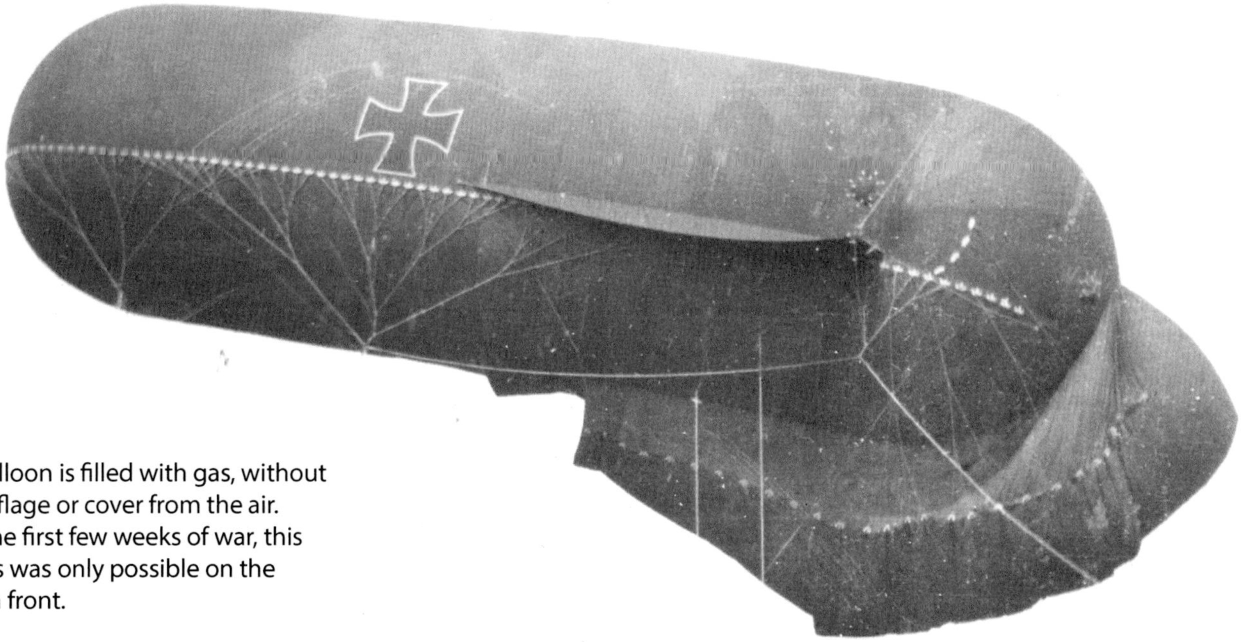

This balloon is filled with gas, without camouflage or cover from the air. After the first few weeks of war, this process was only possible on the Eastern front.

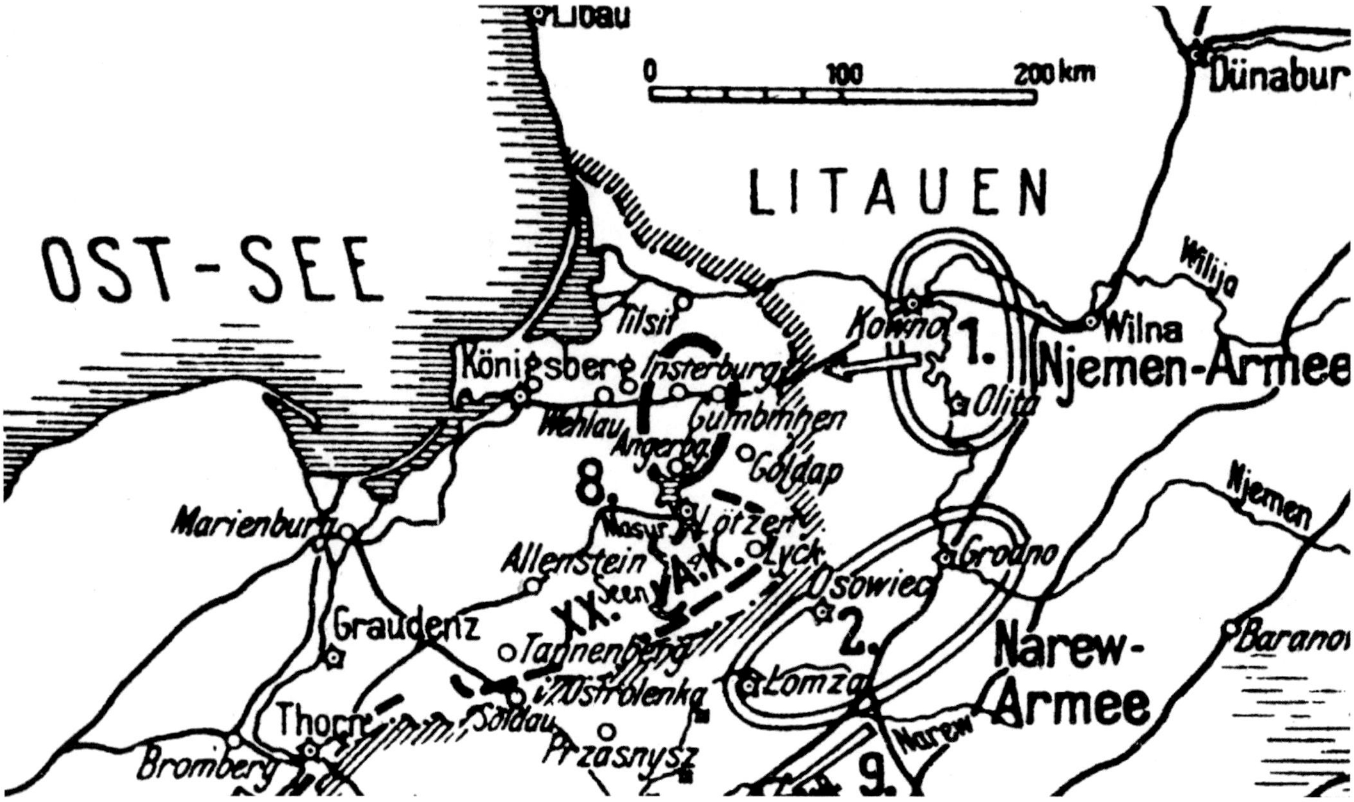

Above: Deployment of the 8th Army in East Prussia, August 1914.

cannon batteries were operating from an open field position, launching cannons towards the west. An infantry batallion gathered for Sunday worship nearby. The 15 FLA 1 horse drawn carriages stood closely together behind the artillery, and the horses had been boarded in a nearby town. Just like in times of peace, these activities took place in the open fields without any aircraft protection. Suddenly, an enemy aircraft approached, dropping a smoke bomb overhead the balloon without being shot at by German troops. The French aircraft circled the balloon several times before returning to its own front. Preparations were underway immediately to move the balloon laterally when it was hit by the French artillery's first granade. Shortly after, more granades followed; thereafter, the enemy aircraft returned, dropping smoke bombs overhead the balloon as well as the German cannons. Hastily, the FLA 1 ground crew moved the winch wagon and the attached balloon 1500 across country. The French aircraft pursued the balloon, continuing to drop incendiary bombs. In the meantime, the French artillery had begun shooting at German cannons, leaving the balloon intact.[16]

Gefreiter (private) Wilhelm Schmidt of the Feldartillerie-Regiments König Karl 1. Württembergisches Nr. 13 took part in the battle of Longwy on August 23, 1914:[17]

"It was a foggy morning; the poor visibility made it impossible to set up the cannons. Around 09:15 we moved into position near Baranzy and were immediately fired at by French artillery. We were not able to see the shooters. A tethered balloon came to our aid. An observer communicated the location of our enemies using a telephone line. The enemy was east of Signeulx. The 5th and 6th batteries immediately engaged the enemy which were annihilated by 11:00. Lateron, we walked over to the French trench where we discovered a 7.5cm cannon as well as a few horses. Shortly after, our balloon was shot at by French artillery fire, forcing us to move our artillery position at a safe distance."

Contrary to the incidiary bombs dropped by the French aircraft which were rarely a threat to the balloons, fléchettes perforated the balloon's gasbag, which at the very least caused severe damage to the balloon. On September 1, 1914, Lt. Albert Moris of the MF 8 Escadrille sighted a German balloon[18] which conducted reconnaissance for the Bavarian 1. Reservekorps near Arracourt in the vicinity of Lunéville. Moris flew overhead the balloon in the Maurice Farman MF VII (No. 123), dropping a round of fléchettes, killing observer Oblt. Albrecht Oldenbourg and resulting in the crash of the balloon. Oblt. Oldenbourg was part of the 21. Inf. Rgt. *Großherzog Friedrich Franz IV. von Mecklenburg Schwerin* before he transferred to the Ballontruppe in 1912.[19]

The Feldluftschiffer-Abteilung 3 took part in the battle of Marne on September 4. It was directly assigned to Gardekorps (2nd and 3rd Army) and utilized near Epernay. It only executed a handful of reconnaissance flights. After the German troops retreated, it remained in the Champagne region from September 14 in the 2nd Army's trench warfare.

The Feldluftschiffer-Truppe on the Eastern Front

Only the 8th Feldluftschiffer-Abteilung transferred to the Eastern Front along with the 8th Army in August of 1914. If necessary, Generaloberst v. Prittwitz, Commander of the 8th Army, had the authority to request the support of Festungs-Luftschiffertrupps 1, 5, 23, 24, and 26. In total, he commanded 6 balloons. Initially, only FLA 8 (under the command of Hptm. Schellbach) was utilized at the Front. In addition to one Parseval-Sigsfeld balloon, Hptm. Schellbach also had two Kugelballone at his command. On August 14, FLA 8 moved into position near Nordenburg in support of I. Reserve-Korps. Reserve-Korps generally did not include any Feldflieger-Abteilungen which is why they relied upon the support of balloons for reconnaissance reports. In the ensuing battles near Gumbinnen, Schellbach commanded the I. Reserve-Korps' balloon to advance in the front lines. In this way, observer Oblt. V. Laffert was able to keep Divisions-Kommandeur, General Kruge, abreast of the troops gaining ground, as well as the position of enemy troops and artillery. On August 18, Laffert spotted the approach of the 1st Russian (Njemen) Army from his balloon's location near Königsfeld. In spite of the German troops' success in the battle of Goldap-Gumbinnen, v. Prittwitz ordered their retreat. Ultimately, this decision led to his replacement by General v. Hindenburg. On August 25, Hindenburg ordered the 8th Army to attack near Tannenberg. According to Oblt. v. Laffert's report from August 26, 1914 at 17:45, the Russian Njemen Army troops retreated from the entire area of Bössau-Sauerbaum.[20] Under Hindenburg's command, both Russian the 1st Njemen and 2nd Narew Armies were defeated near East Prussia; they retreated to a location beyond the Great Masurian Lakes Region.

At the onset of the trench warfare in the fall of 1914, many deficiencies were detected in the equipment of the Feldluftschiffer-Abteilungen. The balloons were not rising high enough, often only 500-600 meters. Additionally, the balloons were unstable in wind speeds greater than 10m/s, resulting in even the most experienced observers suffering from motion sickness. High-performance binoculars and cameras were not available, reducing reconnaissance operations to 4-7km distance. It was not until the end of 1914 that equipment improvements permitted reconnaissance work to a distance of 25km into enemy territory. The only available balloon with a volume of 1000 m^3 was assigned to a Feldluftschiffer-Abteilung at home, reaching a height of up to 1200 meters.

During the early stages of trench warfare, reconnaissance behind enemy lines offered little to spot beyond the usual rail traffic. Artillery fire control and direction by balloons was not discovered to be an advantage until after observers were directly connected with the artillery via a phone line. Artillery ammunition was limited in the German mobilization schedule, and supplies only lasted until October 1914. Ammunition had to be conserved until the industry was fully converted to war production, allowing more ammunition to be delivered to the Front. The significance of the balloons grew further with an increase in artillery utilization on both sides from November 1914. The collaboration between aircraft and balloons also started at the beginning of trench warfare in late fall. Aerial images were frequently delivered to the infantry and artillery with delay, whereas balloon observers were able to deliver information in a much more timely manner using telephone lines. Therefore, aircraft crew often landed near their balloon comrades and directly shared their reconnaissance results regarding camouflaged artillery or gun magazines. Soon, further success stories as a result of the collaboration between aircraft and balloons followed.

As messages from balloon observers were passed directly to artillery and infantry commanders who oversaw Luftschiffer-Abteilungen, these messages remained largely unknown to General and Army high command. Feldluftschiffer-Abteilung liaison officers were not yet known in Army Staff departments. Inspektion des Militär-Luft- und Kraftfahrwesens (General-Major Wilhelm Messing) and Inspekteur der Luftschiffer-Truppen (Oberst Kuno v. Barfus) were based at their official residences in Berlin[21], from where they were unable to represent the interests of their troops at the Front.

Above: A Festungs-Luftschiffertrupps balloon during an exercise. Festungs-Luftschiffertrupp balloons 1, 5, 23, 24, and 26 stood ready to be deployed to the Eastern front if necessary.

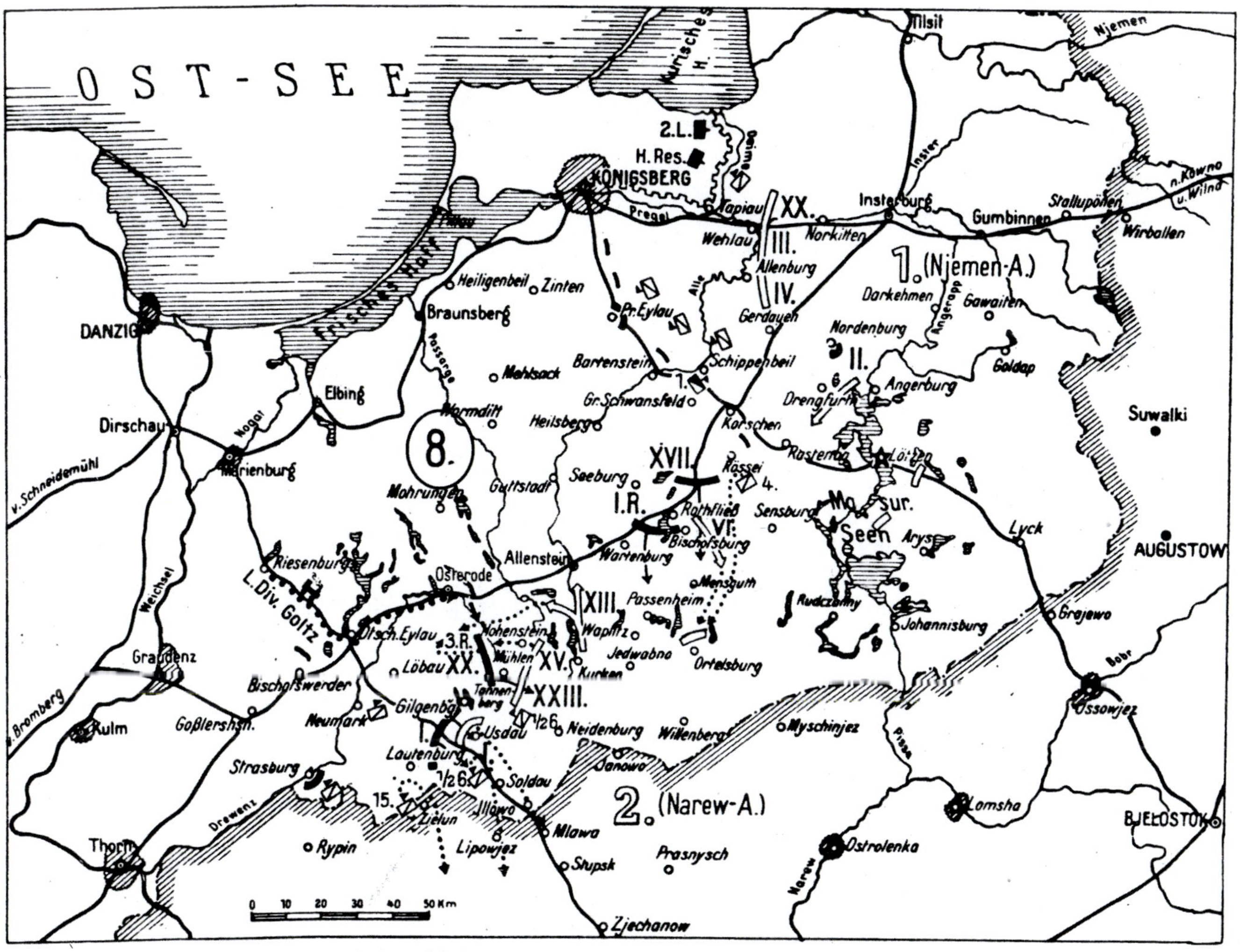

Die Schlacht bei Tannenberg (Lage am 27. August)

Above: Map of the battle at Tannenberg on August 27, 1914.

Endnotes

1 Rangliste der Königlich Preußischen Armee für 1914, Berlin 1914.
2 Bayerisches Kriegsarchiv: *Das Bayernbuch vom Weltkriege 1914-1918*, Stuttgart 1930.
3 Festungs-Luftschiffer-Trupp Germersheim ab dem 24. August 1914.
4 Reichsarchiv: Kriegsrüstung und Wirtschaft, S. 439, Berlin 1930.
5 RLM: Mobilmachung, *Aufmarsch und erster Einsatz der deutschen Luftstreitkräfte im August 1914*, Berlin 1938.
6 Leipziger Neueste Nachrichten, August 1914.
7 Demory, Jean-Claude: Essor de la Motorisation, Boulogne-Billancourt 2008.
8 T = Thomsen, Chief of staff of the Air Service at the time of introduction.
9 Harbord, James: The American Army in France, o.J.
10 Hoeppner v., Ernst: *Deutschlands Krieg in der Luft*, Leipzig 1936.
11 Seither, Philippe: *Guerre Aérienne dans le ciel de Haute Alsace*, Balschwiller 2014.
12 Bayerisches Kriegsarchiv: Die Bayern im Grossen Kriege 1914-1918, München 1923.
13 Mühsam, Kurt: *Die Kämpfe in der Luft*, Berlin 1915.
14 Erinnerungen von Dr. Helmut Göller, Lindenberg (6. bay. Infanterie-Regiment).
15 Mühsam, Kurt: Die Kämpfe in der Luft, Berlin 1915.
16 Riedmann, Michael: *Das Bayernbuch vom Weltkriege 1914-1918*, FLA 1, Stuttgart 1930.
17 Erinnerungen von Wilhelm Schmidt, Kassel (Feldartillerie-Regiment König Karl 1. Württembergisches Nr. 13).
18 Probably the Bavarian Feldluftschiffer-Abteilung.
19 Méchin, David: *The WWI French Aces Encyclopedia*, Vol. 2, Reno 2021.
20 Reichsarchiv Bd. 19: Tannenberg, Berlin 1927.
21 Rangliste der Königlich Preußischen Armee für 1914, Berlin 1914.

Above: Newspaper image of a German balloon in the Great Masurian Lakes Region.

Above: German balloon in the Vosges region, August 1914. August 1914.

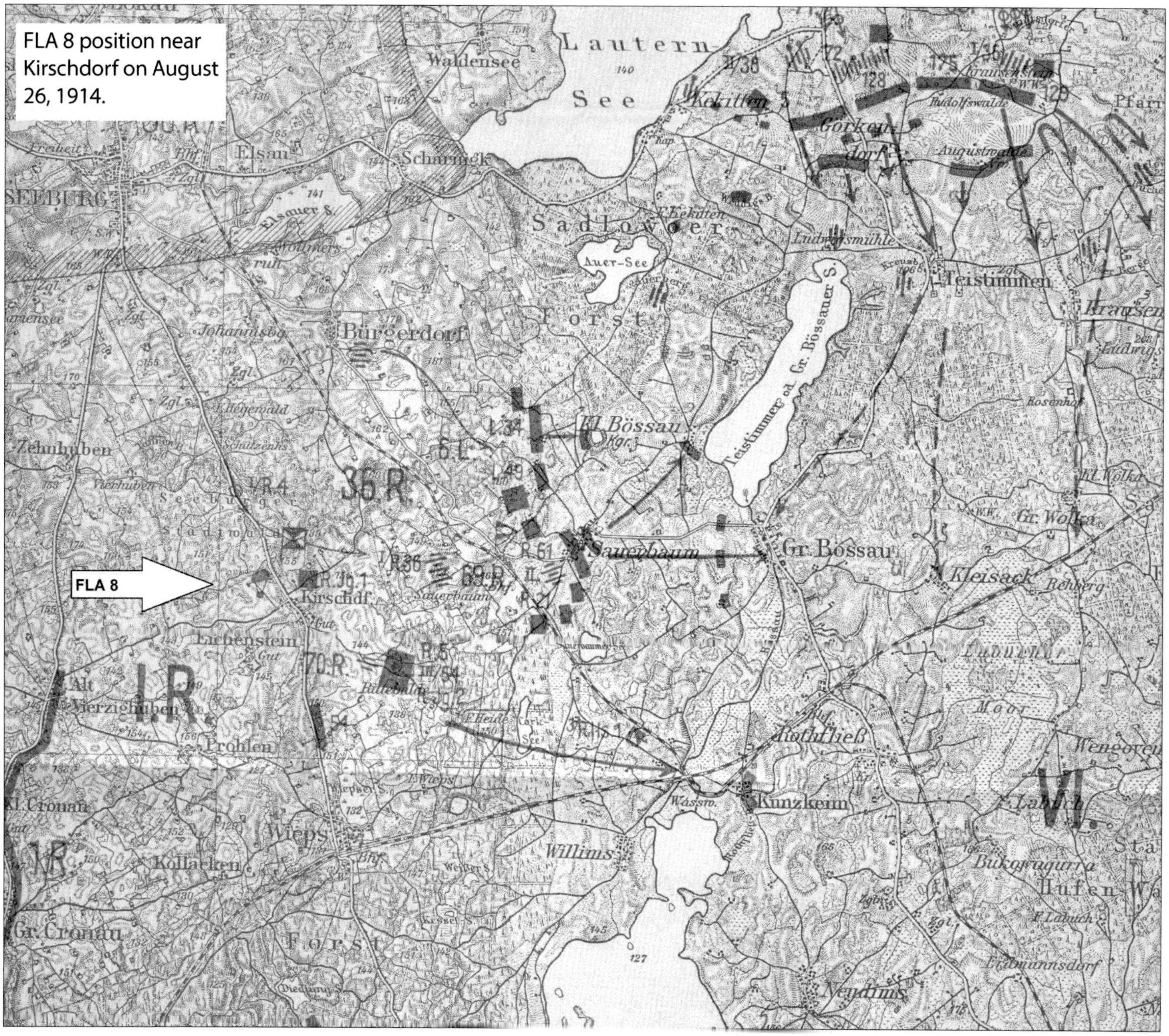

Above: During the first weeks of the war, Bleriot XI aircraft were utilized along with Tauben-aircraft (Etrich-Taube, pigeon shaped aircraft).

Breusch Valley in the Vosges region, fall of 1914.

Battles for Hartmannsweiler and Lingenkopf (Hartmansweiler and Lingen Head, two mountains/ rocky spurs) in the Vosges region continued through the end of the war.

The Year 1915

Above: A truck is used to pull a balloon back to the ground. Castors can be seen between the balloon and the truck.

In February of 1915, nine Luftschiffer-Abteilungen (seven Prussian and two Bavarian FLA) were stationed at the Front. They were assigned technologically improved balloons with a gas content of 800-1000m³ which were able to rise to an altitude of up to 1200 meters and were more stable in windy conditions. The larger size of the balloons, however, was a hindrance when it came to bringing the balloons back down to the ground. It took the crews approximately 1 hour to pull the balloons down from 1000 meters in the air using a hand winch. In order to facilitate the process, horses or vehicles pulled the tether across a set of castors until the balloon reached the surface. In order to accomplish this task, approximately 1500 meters of unobstructed area was necessary for the horses or the vehicle to stretch out the tether.

The open spaces where the balloons were pulled down in turn were more easily spotted by enemy artillery and aircraft. In an attempt to remain undiscovered while preparing for the balloons' ascent, Luftschiffer-Abteilungen made their own motorized winches using captured enemy equipment. During their search for additional engines, the Luftschiffer discovered them at an agricultural division near Sedan, where approximately 50 engines were stored which were then converted into the so-called Sedan winches.

In order to further strengthen and develop the Flieger and Luftschiffertruppe, a cabinet order promoted Lieth-Thomsen's Major Herrmann to the position of Feldflugchef on March 15, 1915. In addition to the pilots and weather service, Herrmann also oversaw the Army's Luftschiffertruppe which included all ront and Heimatverbände (home alliances), and which thereby separated the aviation sector from the Army's transportation units; separating the two was a long overdue step at the outbreak of the war.

As part of the reorganization of the aviation sector, the overseeing troop Feldflugchef required reports about the tasks assigned and carried out by the Luftschiffertruppe. As General- and Armee-Ober-Kommandos issued positive reports following consultation with their respective Front-Kommandeure, the Feldflugchef decided to convert all Festungs-Luftschiffer-Abteilungen into active Luftschiffer-Abteilungen and to deploy them to the front.

Following a three-day barrage fire, British and French troops broke through the German front near Armentières and Arras, as well as the Champagne region between Reims and the Argonne on September 25, 1915. To the north, the British advanced between the channel of La Bassée and Lens near Hulluch. The French took Souchez, Vimy, and Givenchy. In the Champagne region, the area between Aubervive and Tahure, east of Reims, fell into the hands of the French. This Autumn Offensive showed the first signs of a material battle with artillery strikes. The new way of warfare immediately required an increase in balloon units. By the end of 1915, the German Luftschiffertruppe was comprised of 40 units with a total of approximately 80 balloons. Reconnaissance of the enemy hinterland was an important task that was added to the FLA's existing assignment of artillery management. As Feldluftschiffer-Abteilungen stayed mostly on the ground, they were very familiar with the surrounding terrain. Changes in the enemy

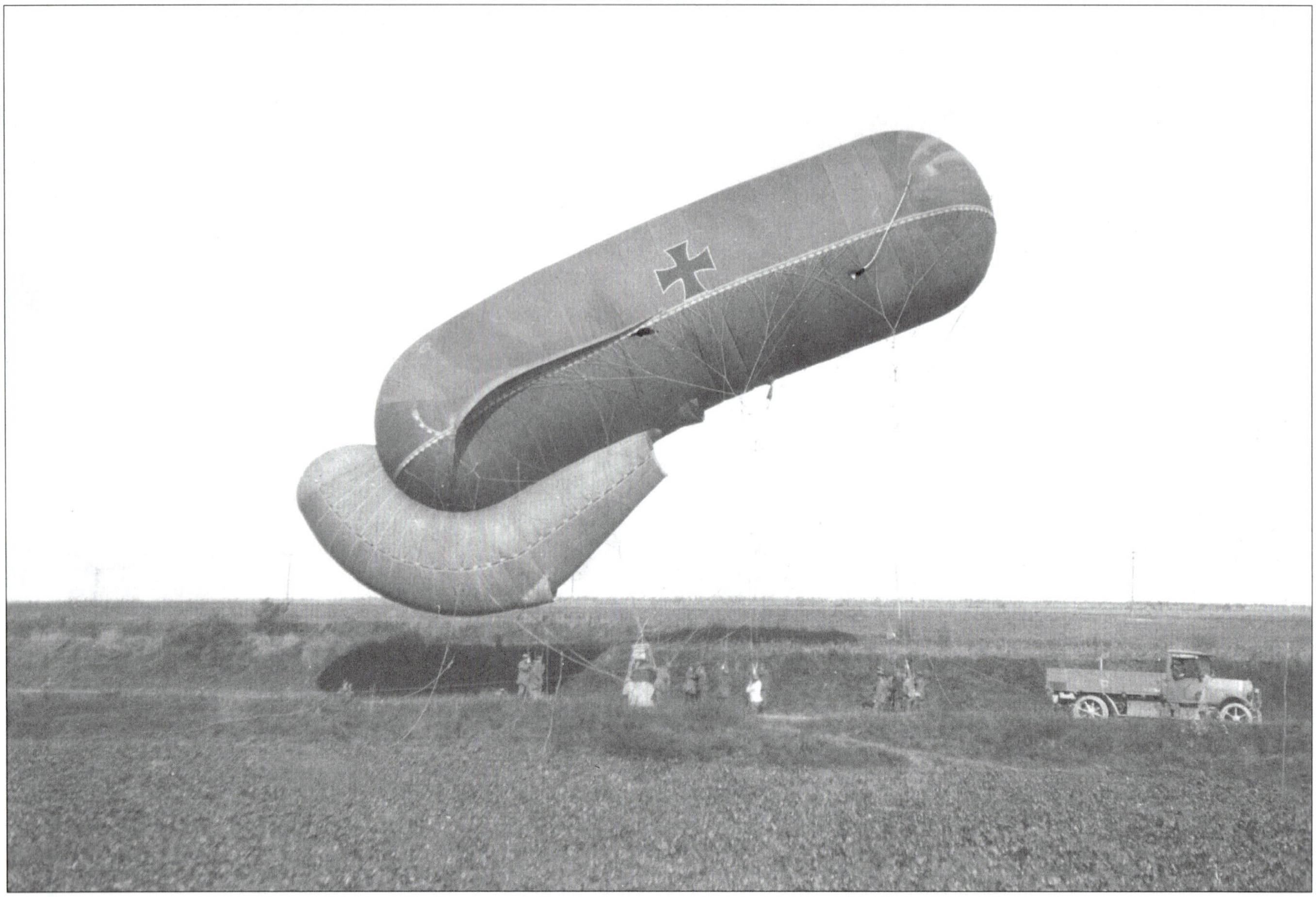

Above: Multipe castors are necessary to pull the balloon to the ground. Ideally, the truck drives along a road during this process.

hinterland, such changes to artillery positions or magazines were therefore spotted in a timely manner. Using 30cm up to 120cm cameras they were able to make detailed spatial image recordings of enemy territories.

After the battle at Tannenberg, the Luftschiffer-Abteilungen of the 8th Army hardly reached any tactical successes at the Eastern front. Poor railway and road conditions prevented regular supply replenishment; furthermore, balloons pulled by horse-drawn vehicles struggled to keep up with the infantry during mobile warfare.

Hindenburg and Ludendorff feared another attack by Russian troops in East Prussia; in early January, they therefore planned a counter attack in Masuria. After prolongued hesitation, General Erich v. Falkenhayn finally agreed to deploy the 10th Army in mid January of 1915.

On February 7th and 8th, the 8th Army attacked; they breached the Russian front and advanced to Augustow. The 10th Army advanced from Tilsit to Grodno and surrounded 100,000 Russian troops. One Festungs-Luftschiffertrupp from Thorn was part of the attack. In May of 1915, the newly deployed 11th German Army and the K.u.K 4th Army breached the Russian front near Gorlice. For this purpose, FLA 25 was assigned to AOK 11 under the command of Hptm. Schlütter. To the north, the 9th Army attacked MIT STOSSRICHTUNG, pushing toward Warsaw. During the attack, the heavy artillery troops were supported by Festungs-Luftschiffer-Abteilungen 24 (Hptm. Martiny) and 26 (Hptm. Conrad). By the end of September, K.u.K and German troops had advanced deep into Russian territory; however, the approaching winter forced them to halt their attacks. The Russian Ossowiec fortress withstood the German attacks and the crossing of the Bobr river could not be completed before the winter.

The Feldluftschiffer-Truppe at the Southeastern Front

On September 6, 1915, Bulgaria joined the Central Powers, a move which was the foundation of the military campaign against Serbia in order to relieve its allied Turkey in their battle on the Straits of the Dardanelles. The attacks began on October 6, 1915, commanded by Generalfeldmarschall von Mackensen who led the 11th Army, 3rd K.u.K Army, and the 1st Bulgarian Army. In Serbia, Luftschiffer-Abteilung 34 rendered good support in spite of a scarce road system. Balloons did not have to fear enemy resistance by aircraft or artillery, allowing them to be pulled almost all the way to the frontline. The FLA succeeded in crossing the Danube river near Semendria and Orsofa using ferries to pull the ballons „hoch" (balloons up in the air). Later, they were barely able to keep up with the infantry on the dirt paths. The balloons did not arrive at the Greek border until the end of March 1916; they were utilized near the Vardar river. Less than 2 km from the front, the balloons rose to heights of up to 1700 meters, yielding favorable reconnaissance results. At times, observers were able to watch ship traffic in the port of Saloniki.

The climate at the Southeastern front consisted of strong temperature variations and had an adverse effect on the personnel and the horses. The southern sun disintegrated the balloon fabric, and rubber bondings dissolved in the heat. Replenishment of supplies stagnated as a result of the distance to Inspektion der Luftschiffer-Truppen in Berlin, gas had to be acquired from Fischamend near Vienna. Finally, the Abteilungen were transferred to Romania. The two Luftschiffer-Abteilungen designated with the support of the Turkish allies did not end up being used in the first place.

As the German Ballontruppe grew in size, inevitably, their losses increased, too. During the first two years of the war, there were three losses of balloons as a result of aircraft attacks, ten as a result of artillery attacks, and 17 due to accidents. Particularly the French *Escadrilles de chasse* of the *Aviation Militaire* whose main task consisted of ground troop support, aggressively attacked the balloons. Therefore, the introduction of Paulus parachutes in the fall of 1915 proved

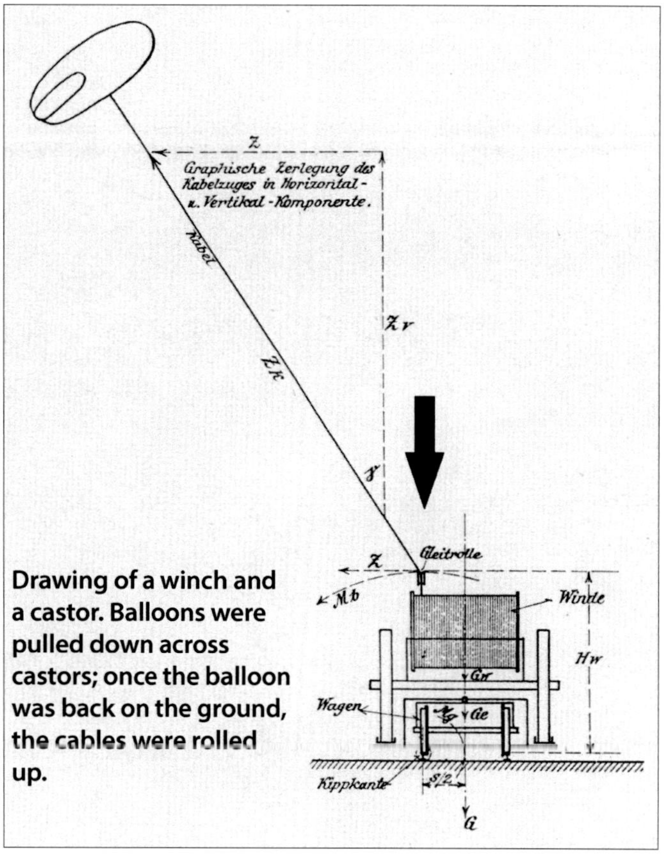

Drawing of a winch and a castor. Balloons were pulled down across castors; once the balloon was back on the ground, the cables were rolled up.

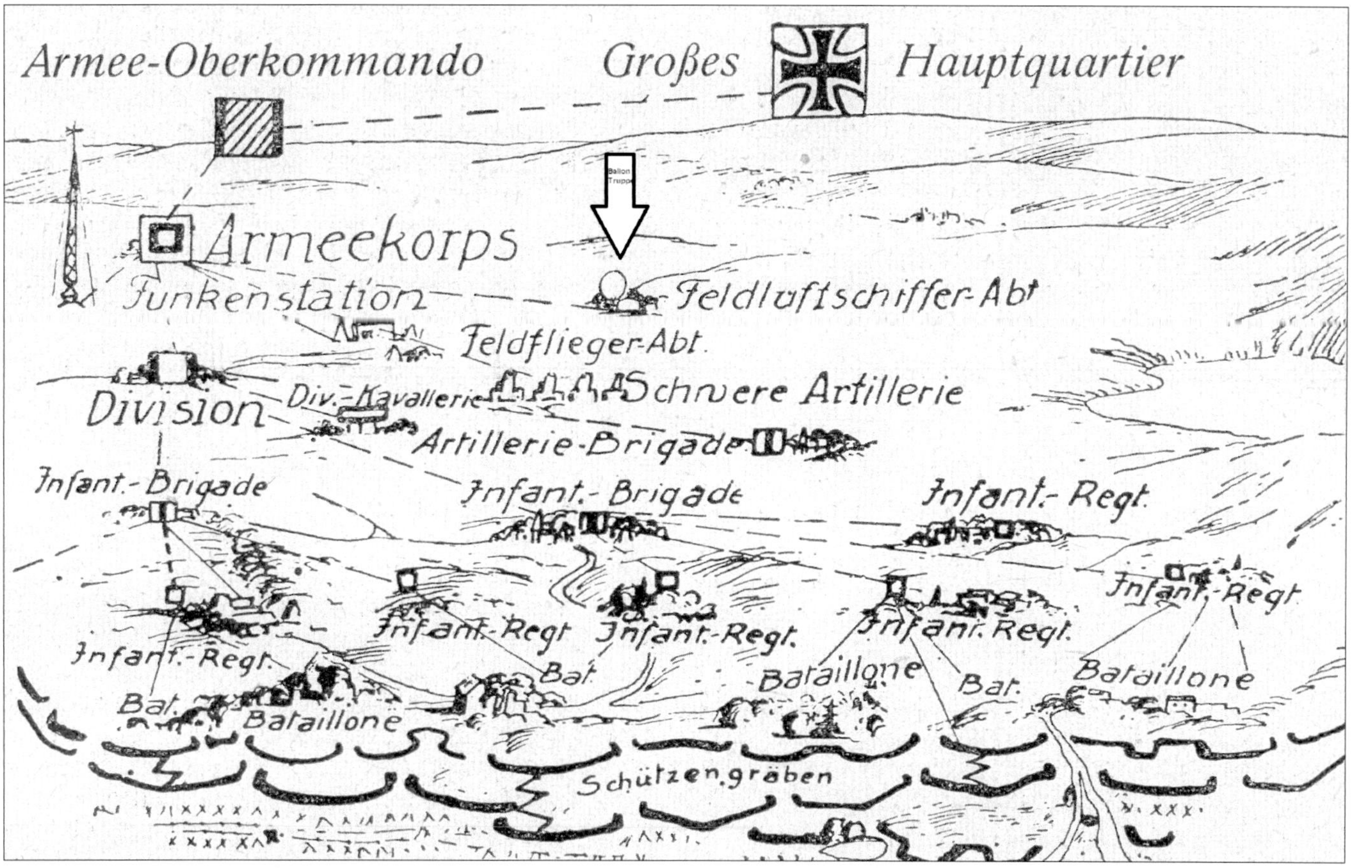

Above: Army command structure in 1915.

Above: A motorized winch, designed by a Luftschiffer-Abteilung. (NARA)

to be the most important technological innovation for the observers within the Ballontruppe. Until then, observers were lost irretrievably as soon as a balloon was shot ablaze. When the easily flammable hydrogen gas in the balloon caught on fire, the balloon immediately exploded, making parachute jumps impossible. The obervers' view upward was blocked by the balloon, forcing them to jump out of the balloon basket within seconds at the order of ground personnel following an enemy aircraft attack. Initially, there were misfortunes due to parachutes failing to open causing the observer to fall to his death. Oftentimes, the reason for the faulty parachutes was the improper packing of the parachute itself. For this reason, many observers tested their parachutes by tossing them from the basket with weights strapped to the parachute itself. Some observers felt restricted in their mobilty by the parachute straps and decided not to wear the lifesaving parachute as a result. Lt. Peter Rieper, an observer with the 7th Feldluftschiffer-Abteilung, frequently stayed in the balloon basket for 15 hours at a time, firing three or four batteries at one time. Rieper wrote the following report:

"Of course, the enemy was aware of our extensive activity; by the fall of 1915, the systemic combat of the balloon was underway. I will admit to feeling quite helpless when I was attacked by four enemy aircraft at the same time in September 1915. Two of them were shooting MGs, the third dropped fuse-bombs, and the fourth aimed at me using fléchettes. We did not yet have parachutes at that time. At most, there was one available parachute for every FLA using three balloons. The only thing I could do was to call for the balloon to be pulled down as quickly as possible."

Above: A Feldluftschiffer-Abteilung 8 balloon in collaboration with an LVG. The balloon was tasked with reconnaissance work for the 1st Bavarian Reserve-Division during the battle of Arras in June 1915.

Above: The Kommandeur and an observer (left) issued a detailed report during an important visitor event at the Bavarian FLA 8.

Above: A Parseval-Sigsfeld balloon in a makeshift hiding spot behind a group of trees.

Facing Page, Below: Artillery and Ballontruppe soldiers engaged in a joint exercise at the Beverloo training area in the fall. A glass of beer was set on the table for the young soldiers. However, neither alcohol nor cigarettes were permitted under the age of 18.

Left: On September 25, 1915, French troops attacked the German front between Reims and the Argonne. Balloon units contributed significantly to the defense.

Above: From 1915, Iron Crosses were painted on German balloon envelopes.

A Ballonabteilungen utilized MG commandeered in battle for defense against enemy aircraft. An example of such an MG was this 7.7mm Fa. Regis Darne & cie., operated by a very young soldier. (NARA)

The Ferme de Navarin Memorial commemorates the battles of the Champagne region.

The Battle of Verdun

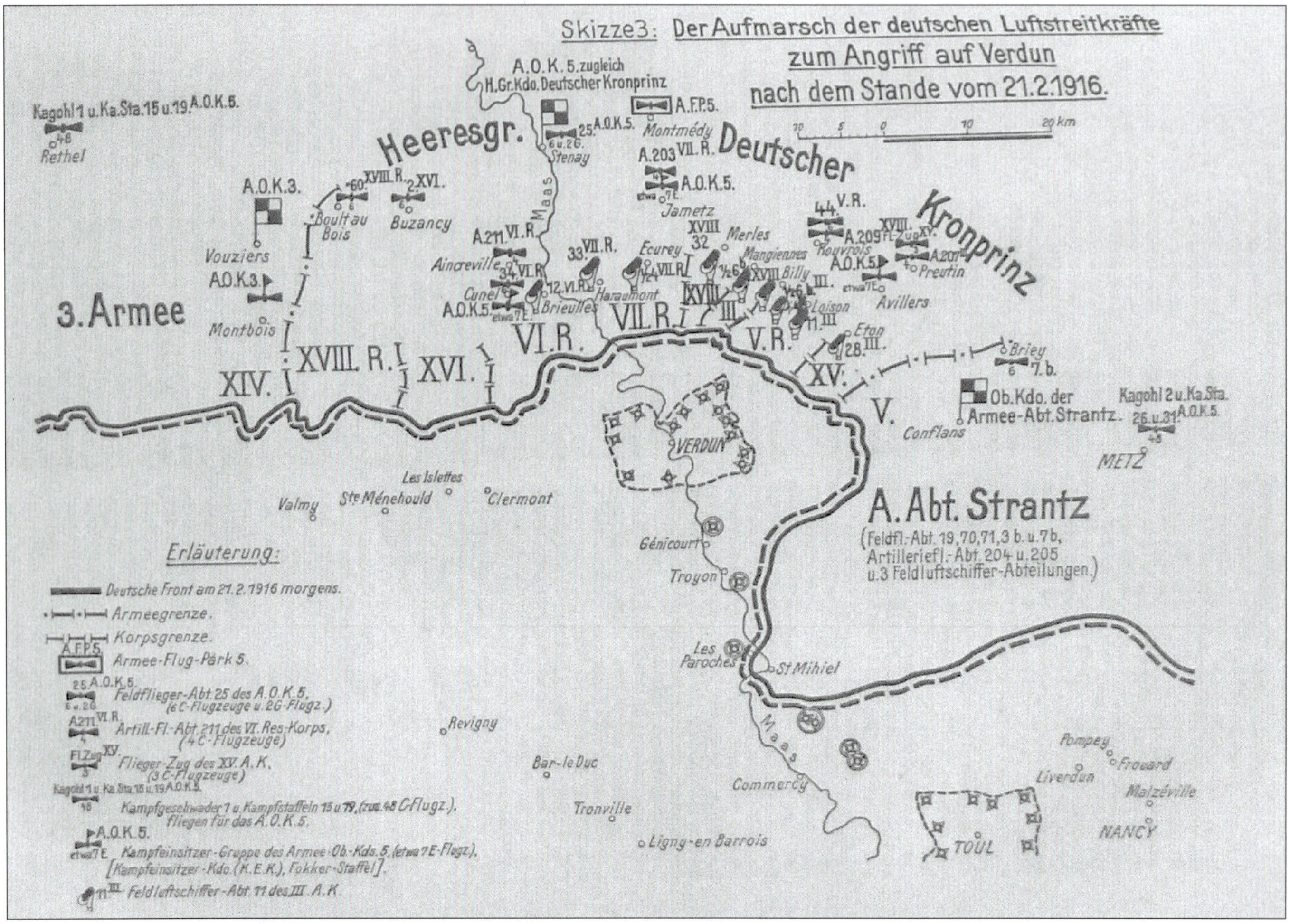

Above: The Flieger and Ballontruppen prepare for the Battle of Verdun in February 1916.

In December of 1915, General Joseph Joffre and Generalstabschef Erich von Falkenhayn decided to bleed their respective enemies to death by means of a powerful battle. However, neither Falkenhayn nor Joffre expected to fight a battle decisive for the outcome of the war. The offensive did not benefit from any underlying strategic maneuvers or genius decisions; material utilization of soldiers, cannons, and grenades were the only thing that counted in this battle. In his 1915 Christmas memorandum, Falkenhayn wrote, "I hope to wear down France using an attack with a restricted number of objectives to such a degree which will leave little hope for them in a military sense." "Falkenhayn chose Verdun as the objective of the attack. From there, strategic rail connections led to Northern France and Belgium, for the defense of which France would sacrifice every last man."[1]

On December 6, 1915, the Entente Generals planned an offensive in Chantilly where the German Army would be bled dry. Joffre chose the Somme for the British-French attack which was to begin in July of 1916. The reason for the chosen time frame was the fact that the 70 British divisions necessary for the attack had not yet arrived in France. By late February, only 40 British divisions had reported for duty at the Western front, three of them Canadian. The remaining 30 divisions were still in training or had yet to be transported to the Western front from Australia, New Zealand, India, and Africa.

Falkenhayn tasked the 5th Army with the German attack near Verdun. The 5th Army was part of crown prince Wilhelm's army group. The first part of the battle would focus on the narrow front between the plains of Meuse and Woevre. The further course of the battle would determine the expansion of the battle area to the Western shore of the river Meuse. The beginning of the battle had to be moved from February 12 to February 21 as a result of reduced visibility due to snow, rain, and fog which prevented the artillery from carrying out its tasks. Secrecy was no longer maintained as on February 10 a deserter revealed the planned attack. Air reconnaissance and espionage provided further information to the French General Staff who managed to

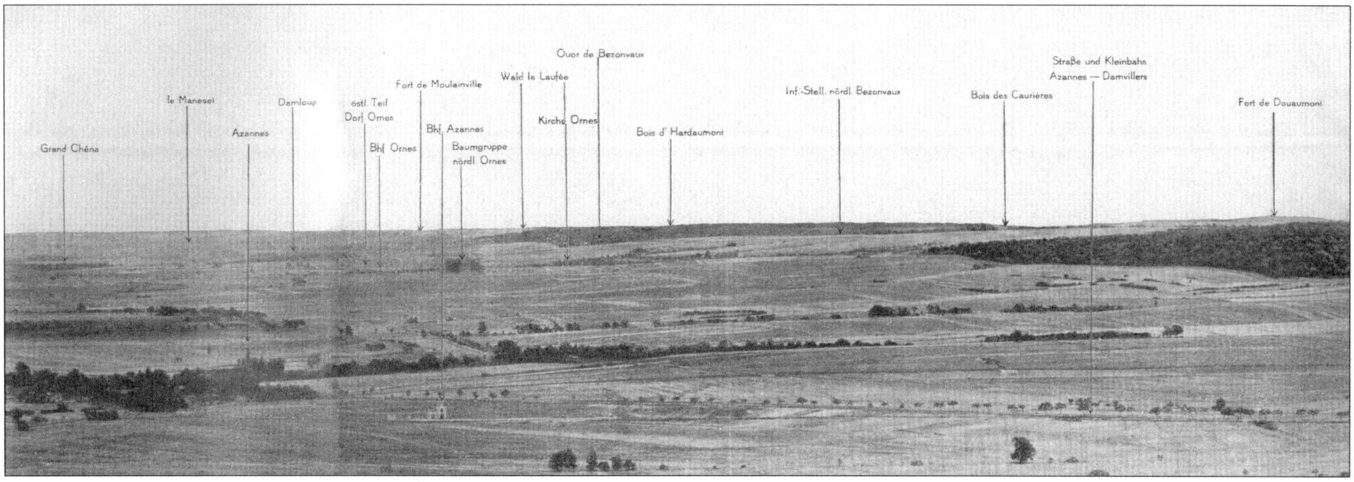

This Page: Aerial images of Verdun as captured from a balloon, 1915.

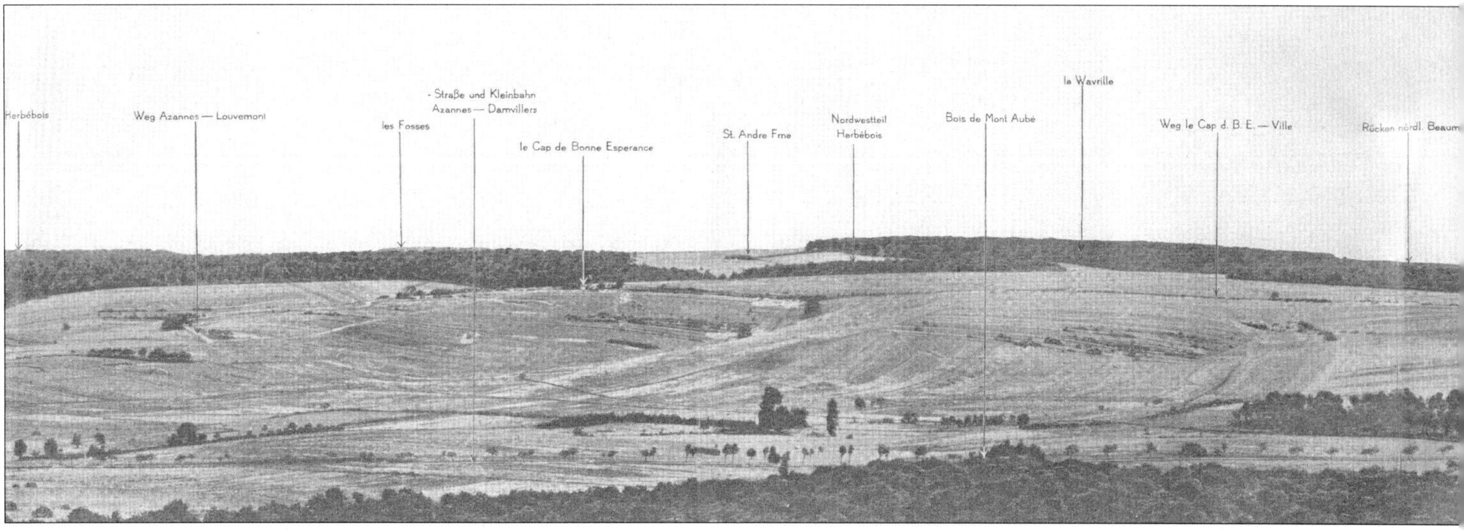

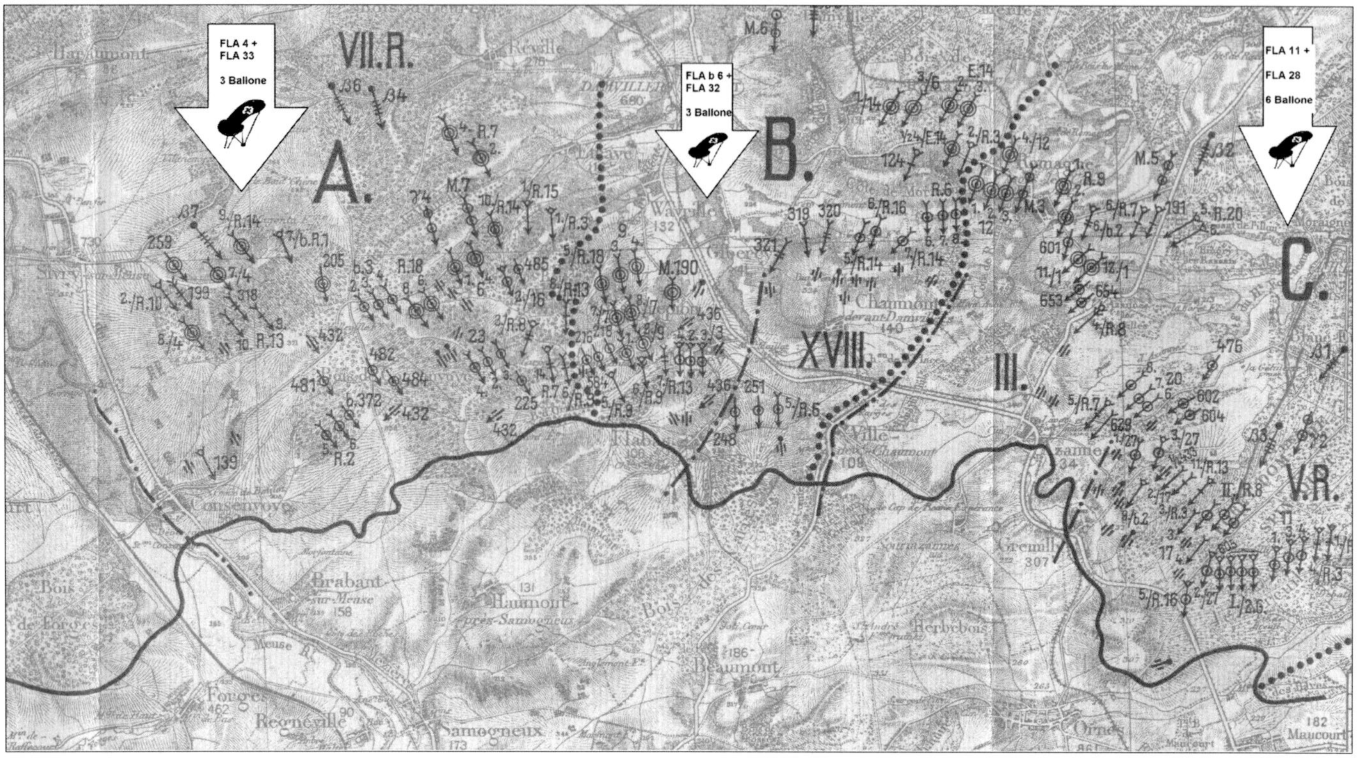

Above: The three groups of attack, including the Ballontruppen, prepare for attack east of the river Meuse on February 21, 1916.

send reinforcement to Verdun just in time.

Three artillery groups were formed along the German attack front, each of which were assigned balloons. In addition, the 3rd Army's OHL temporarily withdrew FLA 11 which was anticipated to return following a successful offensive:[2]

Section A of the VII. Reservekorps including 264 guns. Three balloons supplied by FLA 4 and 33.

Section B of the XVIII. Armee-Korps including 266 guns. Three balloons supplied by FLA b6 and 32.

Section C of the III. Armee-Korps including 320 guns. Six balloons supplied by FLA 11 and 28.

The Fliegertruppe consisted of Artillerie-Flieger-Abteilungen 203, 207, and 209.

Each artillery battery of the three groups received three days' worth of ammunition in the firing position. A Feld-Kanonen-Batterie received 3000 rounds, IFeld-Haubitzen-Batterie received 2100 rounds, heavy Feld-Haubitzen-Batterie received 1200 rounds. German long-range guns of 24cm and 38cm caliber were tasked with the bombardment of forts such as Douaumont or Vaux, as well as the train station, barracks, and Meuse bridges of Verdun.

The FLA 12, supported by 14 balloons, was directly subordinate to the orders of AOK 5 in Stenay. Their balloons were tasked with the leading of long-range guns; they were also to be utilized in potential hotspots during the battle, if necessary. For the first time, a telephonic balloon control network was formed at the Verdun front which led to a collection point in Stenay. A Feldluftschiffer-Depot was used for supply replenishment. The same depot was also utilized to ensure gas supply from Belgium's Luftschiffhäfen. All FLA were supplied continuously with new balloons and parachutes.

On February 21 at 8:12 in Stenay, crown prince Wilhelm ordered the artillery to open fire; storm troopers emerged from the trenches at 17:00. During the first several days, the attacks appeared to go according to plan. On February 25, 1916, Fort Douaumont fell into German hands. The following day, the front lines extended from Vacherauville – Douaumont – Hardaumont. At that time, French artillery opened massive fire on German troops from their positions west of the river Meuse. The defensive began to brace on the eastern shore. On February 28, the German defensive came to a standstill. By the end of February, German storm troops were exhausted; Falkenhayn continued to refuse infantry reinforcement or replacement. Only Fliegertruppen were allowed to receive supplemental units. On March 4, 5 Feldfliegerabteilungen, 4 Artillerie-Flieger-Abteilungen, 26 Pfalz, and Fokker E types, 19 KaGOHL Staffeln, and 8 FLA were utilized at the Verdun front.[3]

On March 6, Falkenhayn finally approved an attack on the western shore of the Meuse, placing the focus on conquering Hill 304 and Mort-Homme (dead man). The attacks of the following weeks resulted in the fall of Cumières, Mort-Homme, and Malancourt; however, the Hills near Esnes and the Fort de Marre remained under French control. The French artillery was positioned there, allowing the Ballonbeobachter of the Compagnien d'Aérostiers an excellent field of vision of the front:

"The French balloon comrades did a good job. Long range

Behind the frontlines, the balloons secretly assumed their ascent positions.

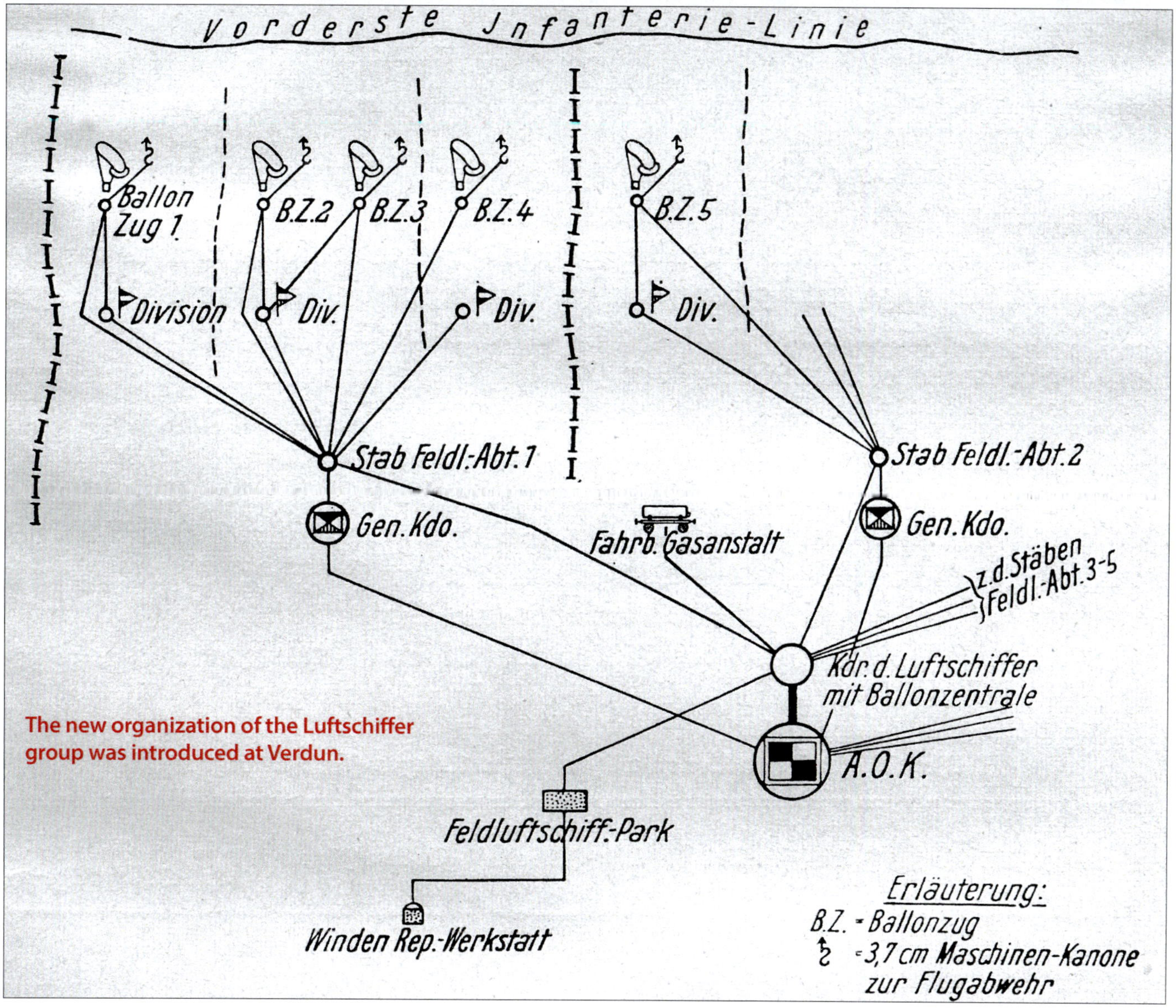

The new organization of the Luftschiffer group was introduced at Verdun.

Left: Every Feldluftschiffer department was issued new equipment before the Battle of Verdun.

Right: A Bavarian balloon prior to ascent.

Above: The artillery prepared for the upcoming Battle of Verdun near Etain.

Above: The Fuss-Art. Rgt 8's 15 cm naval guns began their attack from Cheppy towards Verdun in March 1916. The balloon was observing from its position at Montfaucon.

Above: 21 cm Haubitzen and 15 cm naval guns were set up in the Hinterland of the front in 1916.

Above: For the Battle of Verdun the 5th Army's AOK relocated from Stenay (close to the front) to Etain, east of Verdun.

Above: AOK 5 headquarters location in the Chateau des Tilleuls near Stenay.

artillery troops were shooting down our camps and depots using 22cm grenades all the way to Bezonvaux."[4]

Following is the report of Uffz. Everwien, Artilleriebeobachter of a 21cm Haubitzen-Batterie near the area of the VII. Reservekorps, regarding the impact of a French balloon near Verdun, Hill 378:

"[...] the attack originates from a French long-range battery; it is likely directed by a tethered balloon. We can hear the fire approaching with a howl and the explosions. The heavy 22cm grenades are coming at us at a fast speed as a result of the highest possible increase in powder charge in combination with the length of the gun barrel. A battery salvo is exploding directly in front of us. The Leutnant dropped to the ground immediately; I hover over him. He says, "those bastards, now they've got us." I respond, dusting the dirt off, "there is still time, Herr Leutnant, we have to move out, the balloon is gearing up". His response, "No way! We must stay here. The infantry is not vacating their trenches either." The heavy artillery continues to shoot, the walls of our small, even stand are trembling and vibrating from the heavy impacts. We are lying on the ground, trying to squeeze our heads into the corners. We are yearning for a ceasefire which usually occurs every 30 minutes or so. Then, we sit on the stollen board, fearfully pressed against each other, our heartbeats racing and sitting still and silent. Now, the French are changing their tactics; they had just utilized percussion fuses, and now they are changing the detonators to use a time lag. Our nerves are exposed to a nearly inhumane amount of stress. Heavy grenades, fuses with a time lag are like stollen breakers, intended to crush and trample us. A hunter or sadist must man the French tethered balloon; he wants to destroy us but not before wearing us down. Little does he know he will not have to try; we are already worn down."

It took several hours for a longer ceasefire to occur. Everwien and his still alive comrades took the opportunity and left the observation stand. They managed to escape to the next German position, making their way through completely churted terrain, leaving no trenches recognizable. It took Everwien a few days to recover from a nervous shock which caused him to jump at every loud noise and which triggered strong trembling episodes.

The units of the French Jagdflieger at the Verdun front (*Fortifiée de Verdun* region) consisted of 17 *Escadrilles de Chasse* which had been equipped with modern Nieuport

French aerial image of Fort Douaumont in May 1916.

Aerial image of Fort Douaumont during battle.

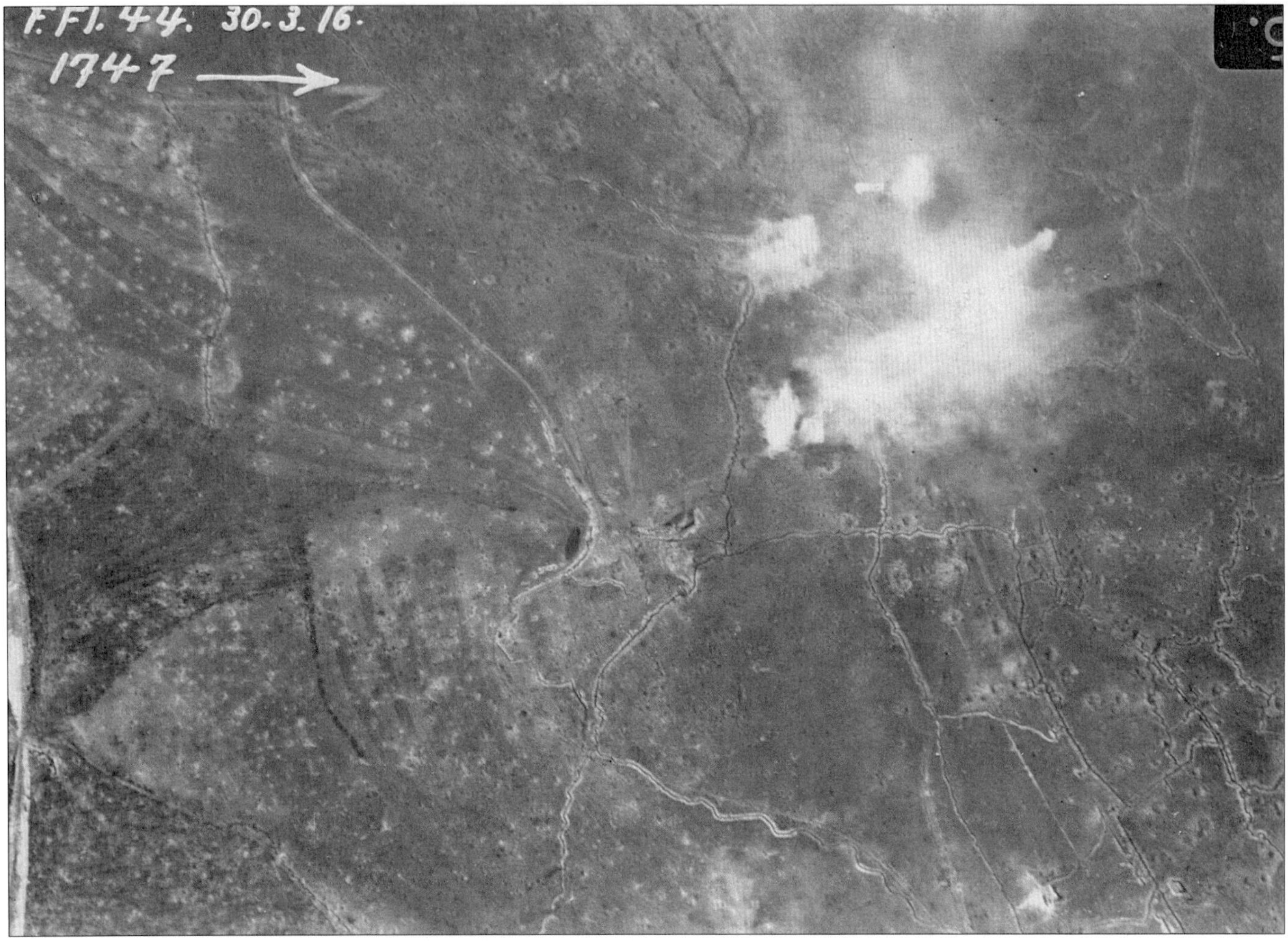

Above: FA 44 aerial image of the Cailette forest, March 30, 1916.

11/16 and Morane-Saulnier L/LA from mid-March 1916. They soon proved their superiority over the German E types. On April 12, the AOK 5 ordered an air barrier over the entire Verdun front; approximately 150 C aircraft supplied by the four KaGOHL flew along the front to support the air barrier. Their primary task of dropping bombs on enemy army encampments, airports, and railway installations lost importance.

Despite all setbacks, Falkenhayn continued the offensive. On April 28, 1916, Général Phillipe Pétain assumed supreme command of the Heeresgruppe Mitte (2ème, 3ème, 4ème, and 5ème Armee); simultaneously, Général Robert Nivelle took command of the 2ème Armee, giving him responsibility of the entire occupied section of Verdun. Both commanders considered the recapture of Fort Douaumont to be their primary goal. This cornerstone of the defense of Verdun

Left: The superiority of the French Escadrilles caused heavy losses in the German Fliegertruppe and balloons. Fokker E aircraft rarely accompanied German reconnaissance aircraft.

Above: The few Fokker E aircraft were technologically inferior to the French Nieuport aircraft.

had fallen into German hands, on February 25, 1916; to the French, this defeat surely felt like a great loss of face.

On May 5, 1916, the Bavarian FLA 6 led artillery attacks on Hill 304 and Mort Homme from their ascent location near Cierges. Observer Vzfw. Reifert had a good view of the target area as the weather was fair and the wind was calm. In the evening, Reifert spotted dark clouds on the horizon; soon after, the first wind gusts began to shake the balloon. As the storm approached, Reifert requested for the balloon to be lowered. The Ballonzug-Führer did not order for the lowering of the balloon until the last minute. The winch had lowered the balloon down to a height of 300 meters when a large gust pushed it to the ground and immediately carried it back up into the air. Gefreiter Josef Eberl described the events:[5]

"A noise and a tug, and the balloon was yanked away with its occupant. The gust was strong enough to overturn the 450 kg winch which had been anchored eightfold, tearing apart the 8mm steel cable. Vzfw. Reifert immediately threw down his carabiner and chart tray as we watched the balloon ascend into the clouds and reemerge, nearly colliding with the ground. Suddenly, a balloon which had been torn away by the French appeared overhead our place of ascent. Several more followed until the number had reached one dozen. While the French colleagues had jumped out of their balloons with their parachutes prior to reaching the front, Reifert was unable to leave his balloon basket. The parachute which had recently arrived had been tested using a dummy that morning but had not yet been reattached to the basket.

As the wind was blowing toward our base, we hoped to be able to salvage our balloon. The same evening, we learned that Vzfw. Reifert had left his balloon near Bantheville. While he was unable to hold on to his balloon, he secured two French balloons using a fleet convoy. The next morning, we received word from the Flieger-Abteilung in Stenay that our balloon had landed in their location. We picked up our balloon and the two French balloons in Stenay later that day. We later learned that an enemy balloon carrying observers had drifted all the way to Gdansk on the Baltic."

At the VI. Reserve-Korps General Command in romagne-sous-Montfaucon, Oberstleutnant Rödiger had just had some coffee in the garden when the thunderstorm began. They immediately sought shelter and soon after, discovered multiple balloons drifting in the sky:[6]

"We counted 22 tethered balloons. While the Germans attempted to lower their balloon by releasing gas, the French chose a different tactic to escape captivity. They tried to maintain altitude in hopes of landing behind the front lines again. Their attempts were fruitless, however; one of them touched down near Aprémont, the other in the vicinity of Montmédy. Some of them drifted as far as Aachen, Trier, and Halle on the river Saale."

The storm surprised 28 observers in 21 French balloons; the balloons drifted towards the east. 12 observers were able to jump to safety using parachutes, landing before the front lines, while 11 were captured and 5 others were killed.[7]

On May 22, Fort Douaumont was to be reconquered by French troops. In addition to the standard, extensive use

German troops conquered the prestigious Fort Douaumont on February 25, 1916. Its two turrets face towards Bezonvaux and the Caurrest forest.

Aerial image of Fort Douaumont, position of the Artillerie-Flieger-Abteilung 203 which had been conquered by the Germans on February 25, 1916.

Above: Fort Douaumont after heavy battles in 1916.

of artillery, this large-scale attack was to utilize the French Fliegertruppe which was tasked with concentrated and offensive charges on German balloons. It was imperative that the eyes of the German artillery along with the balloons of Feldluftschiffer-Abteilungen 4, 11, 12, 28, 32, 33, as well as the Bavarian FLA 6[8] were eliminated before the storming by the French infantry (5ème Division). Le Prieur rockets were to be utilized for the first time in this attack. They were named after their inventor, Marine Capitaine Yves Le Prieur. The rockets were attached to the outermost wing struts, often three or four of them above one another on each side. The pilot would aim the entire aircraft at the enemy balloon and then launch all the rockets which would then burn the entire balloon casing, igniting the explosive gas. In practice, the firing of the rockets was a spectacular sight. On May 4, 1916, Sergent Joseph Henri Guiguet successfully presented this new weapon to a high-ranking commission in Le Bourget; President Raymond Poincaré was among the audience.[9]

At the Lemmes airport, southeast of Verdun, Capitaine Le Prieur gathered eight volunteer pilots from various Escadrilles for a mission. Capitaine Louis Robert de Beauchamp and Lt. De Boutiny (N23), Lt. Jean Chaput (N31), Lt. André Dubois de Gennes and Adjutant Barrault (N57), Sous-Lt. Charles Nungesser and Adjutant Henri Réservat (N65), as well as Sergent Guiguet (N95).[10]

On the morning of May 22, 1916, the eight Nieuport 16 departed Lemmes toward the German front. Weather conditions were clear, making the German balloons easily recognizable from a great distance. Joseph Henri Guiguet succeeded in hitting a German balloon in the vicinity of Sivry-sur-Meuse using a Le Prieur rocket; the ensuing explosion killed observer Oblt. Friedrich von Zanthier (FLA 12). Despite heavy defensive fire from balloon guns and MG, Joseph Guiguet managed to safely return to Lemmes in his Nieuport 16 (N976). FLA 12 Paramedic Hohmeier wrote the following report on the attack:

"On May 22, 1916, Oblt. V. Zanthier, Feldluftschiffer-Abtlg. 12, departed in balloon No. III from the ascent point south of Sivry a.d. Meuse to observe the progress of the attack on the Fort of Verdun. The weather was quite clear with a few isolated, small clouds. The balloon ascended to an altitude of approximately 1200 m. At about 8:30, Oblt. von Zanthier passed the following message to the ascent point via telephone line: "My neighbor balloon is in flames!", and immediately after, "My balloon is on fire!" Observers could spot the burning balloon next to Oblt. von Zanthiers' from the ascent point but paid no attention to their own balloon. Only when the second message was broadcast, carbine shots were heard in the air, fired by Oblt. von Zanthier at an aircraft which was pulling off. Shortly thereafter, a smoke cloud rose from the balloon; we watched Oblt. von Zanthier jump from the balloon basket wearing his parachute. The parachute opened and von Zanthier slowly glided towards the ground. When he was approximately 400 m from

Above: There is a German and French artillery exhibit at the Fort Douaumont museum.

Above: Die FLA 6 was positioned near Cierges along the road from Cierges-Very.

Above: In the early morning hours, FLA 6 troops prepare the balloon for ascent.

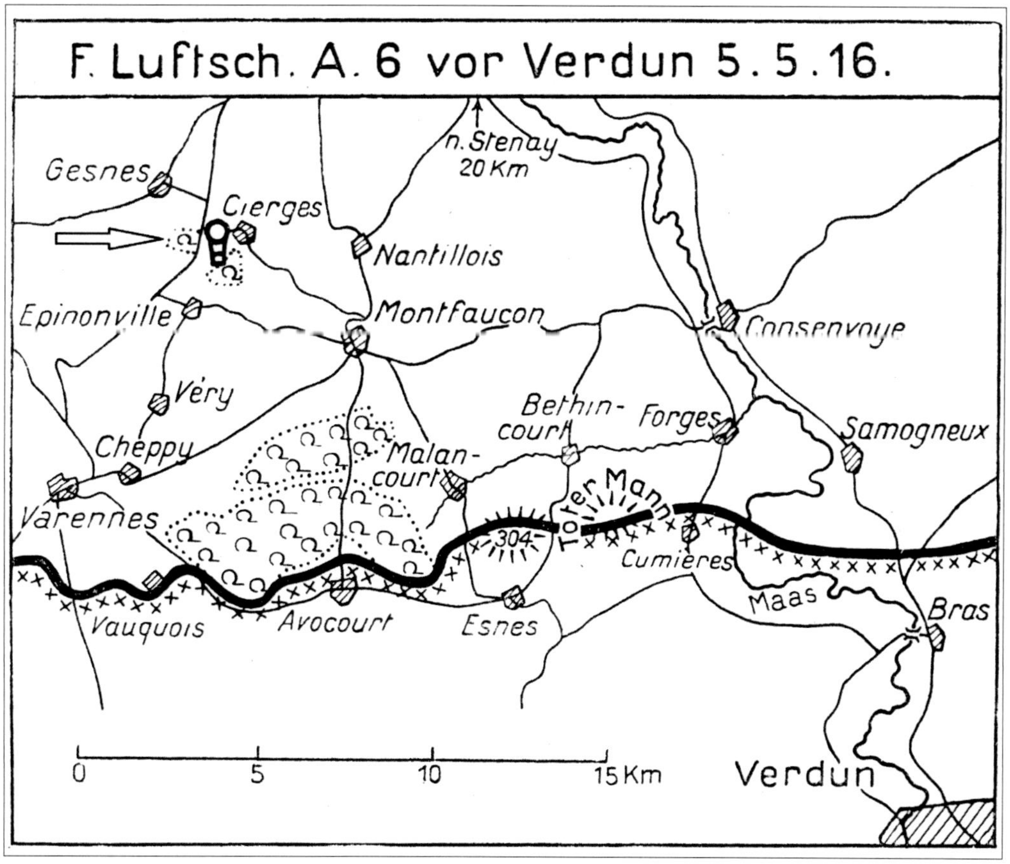

Right: On May 5, 1916, FLA 6 led their artillery fire on Hill 304 – Mort Homme from their ascent point near Cierges.

Above & Below: Hill 304 and Mort Homme were part of the FLA 6 observation sector in May of 1916.

the ground, the burning balloon suddenly came crashing down, colliding with his parachute and pulling him to the ground. Oblt. v. Zanthier's burned, mutilated body was later discovered along with the shattered balloon basket and fragments of equipment parts."

The body was transferred to the cemetery in Sivry where a memorial service was held for the fallen observer on May 24. Division Kommandeur Rittmeister v. Laffert and all officers and crews were present. In the evening, Oblt. v. Zanthier's body was transferred to Metz.[11]

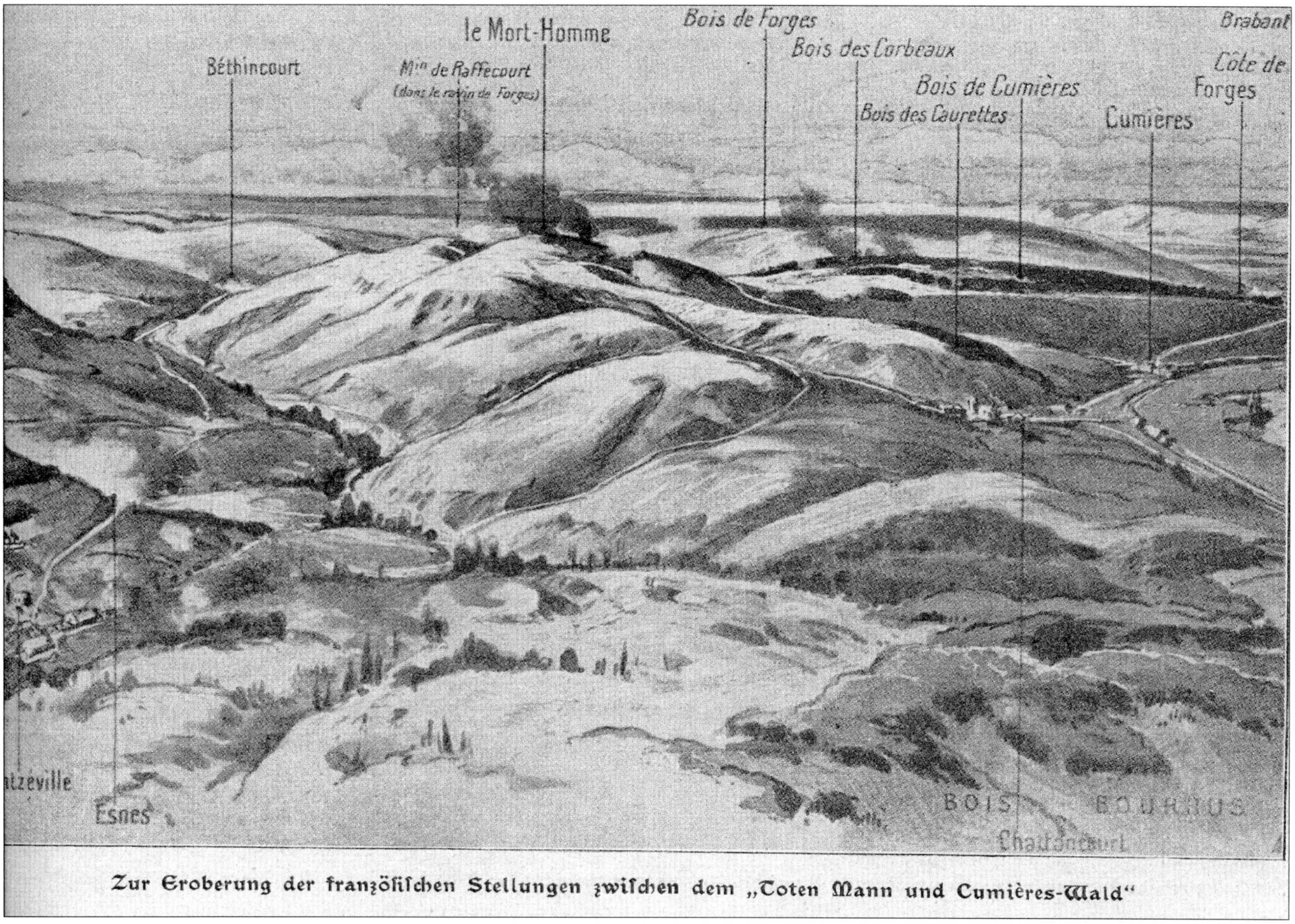

Zur Eroberung der französischen Stellungen zwischen dem „Toten Mann und Cumières-Wald"

Above: Overview of the embattled hills of Verdun.

Above: FAR 99 officers near Verdun; 2nd from right is Oblt. Linneborn.

Above: FAR 99 firing position. Enemy reconnaissance from the ground was nearly impossible.

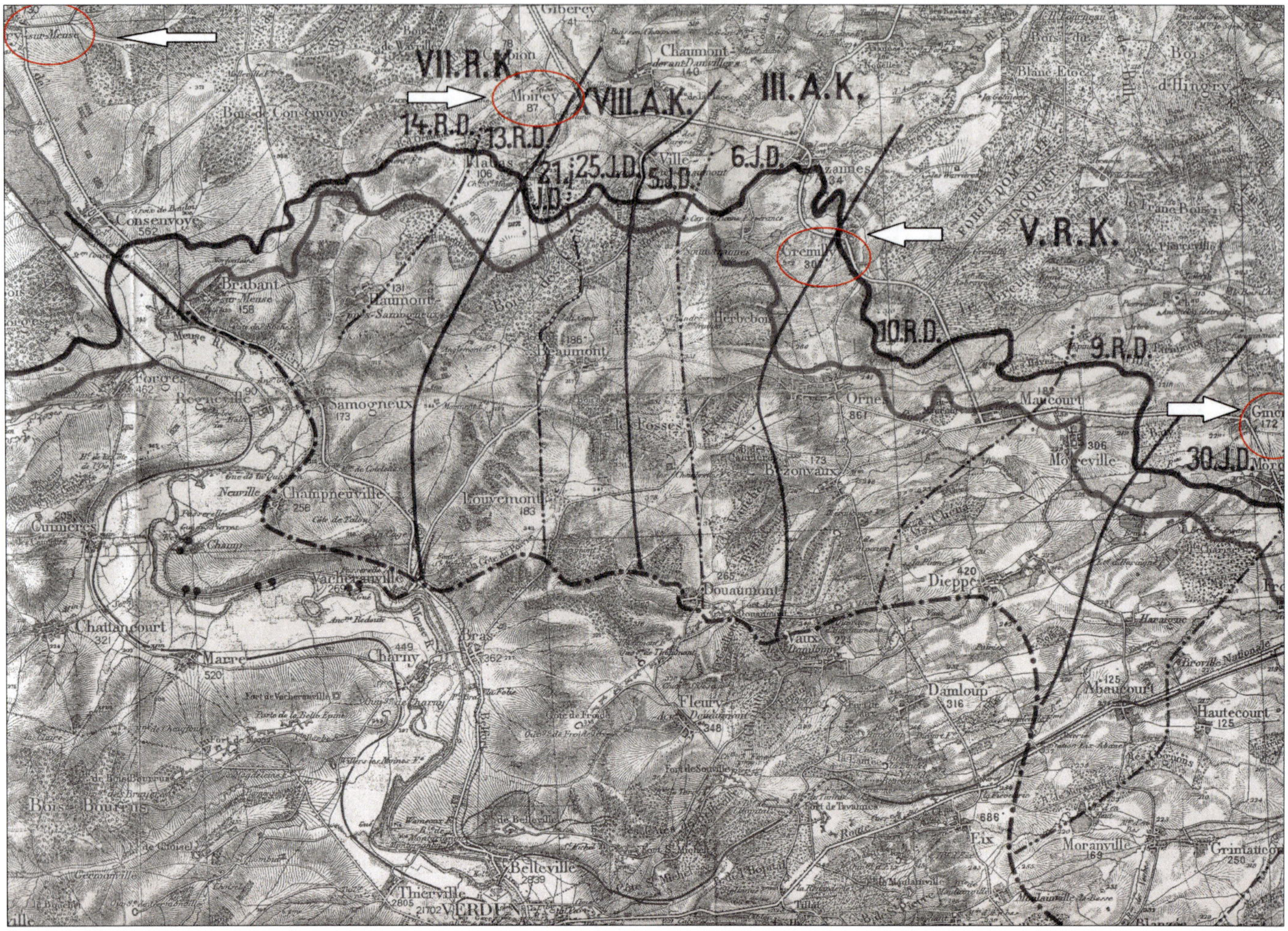

Above: Map of Verdun. Arrows indicate the successful balloon attacks on May 22, 1916.

Left: French soldiers at Fort Vaux in June 1916. The remains of the fort are barely recognizable.

Right & Below: Capt. Le Prieur first introduced his rockets on May 4, 1916. Sgt. Guiguet successfully bombarded a captured German balloon during a test.

Above: Sous-Lieutenant Charles Nungesser.

Between Moirey and Gremilly, De Beauchamp, Chaput, and De Gennes had their successes, each shooting down a balloon. De Boutiny was unable to release his rockets due to a technical malfunction. Nungesser and Réservat each shot ablaze a German balloon of the 4th and 11th Feldluftschiffer-Abteilung in the 50th Infanterie-Division sector (V. Reserve-Korps) north of Gincrey. Adjudant Barrault's rockets missed their target.

After shooting down the balloons, Nungesser and Adjudant Réservat were attacked by three German KaGOHL 2 patrol aircraft. Réservat had to make an emergency landing after a short aerial battle; he ended up in

Above: Souvenir photo in front of Adjutant Henri Réservat's Nieuport 16.

Above: Nieuport 16 N959 bears a white, curved letter R on the side of the fuselage.

Above: Haptmann von Detten was killed on 21 June 1916 near Genicourt.

Above: Oblt. Friedrich von Zanthier's final resting place at the cemetery in Metz.

Left: The Bavarian FLA 6's balloon is pulled to the ground.

Below: Balloon aerial image of a destroyed village in the vicinity of Verdun.

Above: French artillery, led by observers of the Compagnie d'Aérostiers, bombarded German positions from the hills near Esnes and Fort de Marre.

Balloon ascending at Gremilly in late summer 1916; the balloon hangar can be seen in the background.

Above: At the end of August, the Bavarian FLA 8 left Verdun for the Somme region. Comrades and the squadron dog watch as Offiziere and Vizefeldwebel play a round of chess in front of the officer's messroom.

German captivity. Oblt. Kuno Linneborn of the 99th Feld-Artillerie-Regiment witnessed the battle; the regiment had been deployed near Bezonvaux- Dieppe on April 14, 1916, in support of the 50th Infanterie Division during the storm of Fort Vaux.[12] In a letter, Oblt. Linneborn described the aerial battle he witnessed on May 22, 1916:[13]

"[…] On May 22 in beautiful weather conditions, a bold French pilot shot five of our tethered balloons ablaze with wing-mounted guns; all the observers perished in the flames. I wonder if this was the revenge for all the French balloons came loose in a sudden thunderstorm a few days ago and drifted towards the east where some of them landed in Central Germany. The French, however, did not anticipate the three German aircraft who suddenly encircled him. Following a violent aerial battle, he was forced to land close by. Of course, I rushed to take photographs of the occurrence. While observing the battle which was carried out at 30-40m, my role was passive and unfortunate as the Germans fired at

Above: The few remaining Fokker E aircraft were still attached to the Flieger-Abteilungen.

Right: German C aircraft, such as the Aviatik C.II pictured here, were significantly inferior to the French Nieuport. Two flexible IMG 08 were attached to rails alongside the forward observer's cockpit for defense purposes.

Above: Mort Homme was treeless after the war.

Above: This French monument at Mort Homme features two German 7.5 cm field guns and serves as a tribute to the bloody battles that took place here.

the French from above, shooting down on him. I was nearly hit, so I ran to seek cover. The Germans, too, landed close to the spruce fighter biplane. Come to find out, one of the German flyers was an old friend of mine from our days at war school, Gerbens. Of course, our reunion was celebrated with due diligence.[…]“

Adj. Réservat was taken captive by artillerists and, on order of the German flyers, was brought to the FA 44 airfield where he spent the rest of the day as a guest before being transported to the prison camp in Germany. On March 19, 1917, Réservat managed to escape German captivity and make his way back to his homeland.

The missing German balloons at Douaumont posed a grave disadvantage to the German defense. While the French artillery positions were recognized and counter struck by balloon observers on May 21, the German artillery remained nearly blind on May 22. The I. Batl. Leib-Grenadier-Regiment 8 and the I.Batl. Inf. Rgt. 52 were forced to forfeit their positions in the vicinity of Douaumont following heavy losses caused by artillery and gas attacks. They retreated to the fort. Shortly after 14:00, the first French soldiers were positioned at the Douaumont and entered the fort. There, they briefly conquered the southwestern and western parts including the all-dominant, southwestern tank turret. However, they were soon driven back out of the fort.

At the end of May, German troops conquered Hill 304. On June 7, the French troops of Fort Vaux surrendered. Every part of the ground had been fought and paid for by thousands of deaths on both sides. This included in increase in losses for German Ballon and Fliegertruppen. Hptm. Günther von Detten (KaGOHL IV, Staffel 23) fell on June 21, 1916, during a KaGOHL 4 bomb attack on an access roadway near Génicourt-sur-Meuse. The well-known pre-war flyer holding FAI license Nr. 234 would not be the last of his kind from the aviation pioneering days. Von Detten's Rumpler C.I crashed in the vicinity of Fort de Génicourt after a brief aerial battle involving Sous-Lieutenant Jean Chaput of the N57 (7th aerial victory)

While no victories were recorded in the battle of Verdun during the month of June, Chief of Staff Erich von Falkenhayn did not abort the attacks. In addition, news of the 2nd army on the Somme were a strong indication of an imminent Allied offensive attack. It was not until July 11 that Falkenhayn ordered strict defensive measures, just a few days after the beginning of the Somme battle.

Meanwhile, the French had transitioned from defense to offense. On September 2, newly appointed Chief of General Staff, Hindenburg, ordered the termination of all attacks. "The world's best two armies met at Verdun. Without much of a noticeable change, the youth of both nations senselessly bled to death." [14]

Following the re-conquest of the prestigious Fort Douaumont on October 24, the French pushed their lines towards Ornes. The battle of Verdun ended in December of 1916 without a clear outcome. France and Germany each recorded grave losses during the battle from which neither side was able to fully recover. The frontlines changed only little until 1918, when larger scale battles were fought following the arrival of US troops.

The Battle of the Somme

Above: A British kite balloon before its ascent.

On June 24, 1916, a never-before seen barrage fire between French and British artillery began the battle of the Somme. In the northern sector of the 2nd German army commanded by General Fritz von Below, the German front was to be forcefully torn apart according to the plans of Generals Joffre and Haig. Artillery preparations were initially set for five days. The British and French had more than 1655 field guns and 1348 heavy artillery at their disposal.[15] The airspace overhead the Somme had been controlled by RFC and *Aviation Militaire* since April 1916. On June 23, AOK 2 issued the following report:

The German artillery is coming under intentional fire using aircraft observation.

Enemy aircraft are circling to the road Arras – Bapaume – Pérronne.

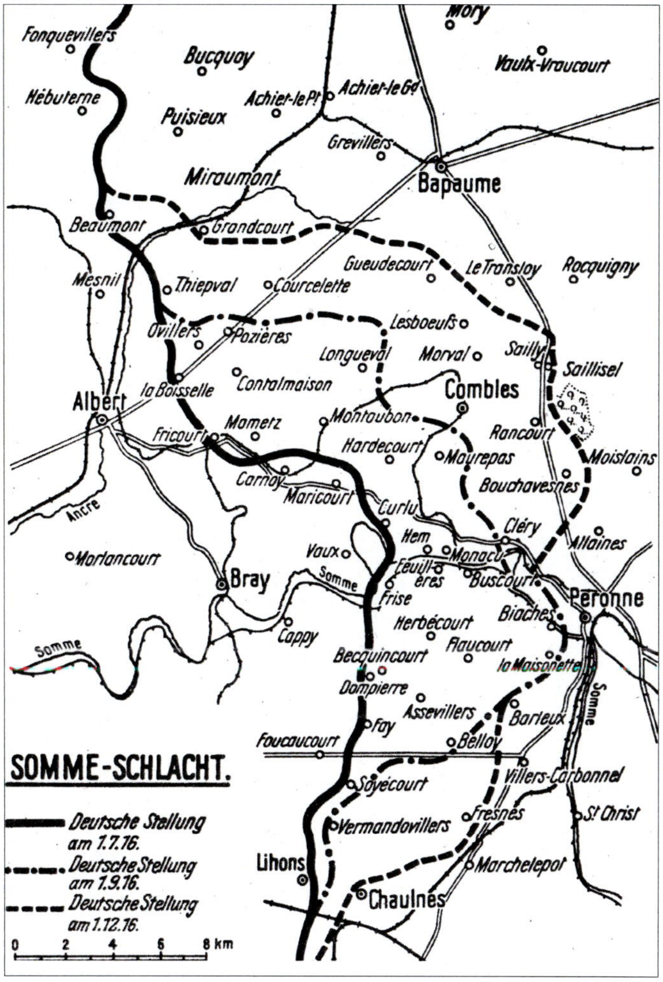

Above: The Somme Front in June 1916.

Above: The British trenches photographed from a balloon in June 1916.

The numbers and quality of our aircraft are no match to the number and quality of their aircraft.

The enemy is in control of the airspace.

An entry in the war diary of the 99th Reserve-Infanterie-Regiment reflected the superiority of the British-French

Above: Balloon of Field Artillery Regiment 46 at rest.

Above: Artillerymen and balloon soldiers on the Somme in May 1916.

Above: A replacement balloon of FLA 2 at Combles in early July 1916.

Above: The German artillery fought practically blind after the loss of their balloons.

Above: Funeral of a member of FLA 2.

aircraft:[16]

"In April of 1916, the regiment set up camp near Thiepval. Enemy aircraft circled over the trenches every day, flying low and attacking us with their MGs. On May 26, Unteroffizier Michalski used a rifle with a telescopic sight to shoot down a British aircraft which made an emergency landing south of Mesnil.[17] At 19:45 on the following day, the artillery succeeded in hitting an enemy aircraft over the trenches.[18] The number of British tethered balloons increased daily, balloon observers practiced with flashing lights without any interruptions. German aircraft were wholly absent."

The French Aviation Militaire consisted of Groupement de Combat de la Somme (GC Somme) using N3, N26, N73, and N103 (founded in April), as well as Groupement de Chasse Cachy (N5, N23, N37, N62, N65, N67, and N69), which were equipped with Nieuport 11, 16, and 17. Their pilots were encouraged to focus their attacks on the German trenches and artillery posts. As a result, German infantry soldiers were left feeling completely defenseless in the face of the enemy aircraft attacks.

Bombers of the Groupe de Bombardement 3 (GB3) with Escadrilles VB107, VB108, and VB109 (Voisin 3,4, and 5), the Groupe de Bombardement 5 (GBM5) with Escadrilles BM117, BM118, BM119, and BM121 (Bréguet-Michelin 4 and 5), and VB101 (Voisin 3, 4, and 5) were exclusively deployed for the Somme attack. It focused directly on the German supply installations and airports.[19] The Royal Flying Corps air forces (4th Brigade) had a total of 167 aircraft at the onset of the battle of the Somme. On April 1, 1916, the number of aircraft in each squadron was increased from 12 to 18. The Ballontruppe consisted of the No. 1 Kite-Balloon-Wing.[20]

At the beginning of the battle, the outnumbered 2nd Armee only counted six Feldflieger[21], three Artillerieflieger-Abteilungen, and 16 Jagdflugzeuge.

On June 25, 1916, British fighter aircraft shot down three FLA 2 balloons (Kdr. Hptm. V. Zech); observers Lt. d.R. Weite and Anton Brinckmann, as well as Vize-Wachtmeister Thomas Auchter, Lang, Schneider, and Fritz Sedlag perished. The balloons were staffed with two observers each and were not equipped with parachutes. The experienced observers had changed from Artillerie-Truppen to Luftschiffer; on that day, they were busy setting up the few German guns for battle

Above: An observer from the Compagnie d'Aérostiers prepares the artillery fire using a map.

Above: Rumpler C.I (431/16) as an infantry aircraft over the Somme.

Above: The hard-fought Beaumont in the north of the summer front, just a few more stones.

Above: Fokker E. III crash-landed in a cornfield on the Somme.

against enemy artillery. Aerial victories were awarded to Lt. Moore, Lt. Latta of No. 1 Sqn., 2Lt. Ball of No. 11 Sqn., and Lt. May of No. 2 Sqn.[22] In the same timeframe, 14 British balloons were stationed at the front which were continuously protected by two aircraft.[23]

According to statements made by deserters and captives, the British-French attacks were set to begin on June 29. However, heavy rain postponed the infantry attacks to July 1. At 06:30, infantry soldiers stormed the remaining German trenches between Beaumont and Clery.

At the onset of the ground offensive, MG and Le Prieur rocket attacks increased on German balloons which had been moved to the front as a replacement. The 16 German fighter aircraft were unable to prevent the attacks, resulting in the balloons falling victim to the French Escadrilles de chasse and the British RFC.

The German anti-aircraft gunners were not a sufficient means of defense.[24] While its numbers could be increased to 100 guns by July 15, flak artillery continued to be outnumbered against the superiority of the Allied Forces.

To facilitate the chain of command, OHL divided the contested area of the Somme on July 19, 1916. The northern sector of the front continued assigned to the newly deployed 1st army under the command of General von Below. The 2nd army was now only responsible for the area south of the river; it was commanded by General Max von Gallwitz

Above: A Fokker E.IV of FA 206 also had to make an emergency landing after a dogfight.

Above: British POWs on the Somme in August 1916. A German soldier also joined the medics.

who was also commander of the army group formed by the two individual armies. More than 30 balloons ascended on the French and British sides of the front every day. The balloons were connected telephonically and uninterruptedly led artillery fire attacks on the few remaining German canons. By August 7, six additional German balloons were

Above: The village of Bouchavesnes in September 1916.

Above: A balloon of Bavarian FLA 8 is prepared for ascent behind a hill.

destroyed, while only a single French balloon was shot down. The observers in the reserve balloons tried their best under the difficult circumstances. Oblt. V. Reckow of the 2nd Nassau Feld-Artillerie-Regiment Nr. 63 reported about the operation of his guns on July 29, 1916:[25]

"Thick fog prevented any observation of the enemy. All telephone line connections with our infantry are shot down and destroyed. Only the rearward monitoring position (balloon) remained intact."

Ultimately, 18 Feldluftschiffer-Abteilungen were deployed to the Somme, utilizing 50 balloons. The Feldflugchef and leader of the army group, Gallwitz, urged the OHL to deploy an even larger number of balloons led by a knowledgeable officer.

From August 24, the British and French resumed their united large-scale attacks on the German front, focusing their efforts on areas near Thiepval and in the vicinity of Soyécourt. Despite mutual grave losses, neither side was

Above: The Somme Front in June 1916.

Right: The Bavarian Feldluftschiffer-Abteilung 8 took up temporary residence in Manancourt Castle Park.

Above & Right: The ground crew of the Bavarian FLA 8 lived in tents. The castle served as a military hospital and was not allowed to be occupied by active troops.

able to record significant special gains until September 12, when the French succeeded in advancing deeply behind German lines. In collaboration with British troops, they were able to further increase their success near Thiepval and Rancourt. The German troops managed to halt the large-scale attacks once again and in late September, the front lines strengthened between Thiepval- Courcelette- Gueaudecourt-Morval-Rancourt-Bouchavesnes, and for the first time in war history, tanks were deployed near Flers. While the British and the French continued their large-scale attacks until the end of the battle of the Somme in November 1916, no significant breakthroughs were reached.

In early September 1916, the Bavarian Feldluftschiffer-Abteilung 8 was transferred to the sector of the 1st army on the Somme. The newly established Jagstaffel 2 arrived on the same day and took their positions near Vélu. After the ascent point had been determined, telephone operators lay the phone lines leading to the artillery batteries whose fire would be directed by balloon observers. Early in the morning on September 2, the balloon ascended to a height of 1600 meters in shallow ground fog conditions. An additional 24 German balloons ascended simultaneously, as well as on the British side. At 06:00, enemy artillery fire opened suddenly.

In September 1916, Bavarian Feldluftschiffer-Abteilung 8 moved to the area of the 1st Army on the Somme. The crews and vehicles were accommodated in a farmhouse, while the balloon was naturally camouflaged in a forest.

The observer immediately reported the enemy guns to his own artillerists; at times, the observer discharged three of his own guns on the enemy. The British responded immediately; the artillery attempted to shoot down the balloon and placed grenades on the ascent point. When their attempts were unsuccessful, the British and French attacked the German balloons from the air. The observer reported the following occurrences:[26]

"Suddenly, I detected a plume of black smoke rise where the German balloon had been located on the far right. The next plume of smoke followed immediately. I watched the observer jump out of the basked and open his parachute. Two observers jumped out of the balloon next to it. The parachute of one of them failed to open and he plummeted until he collided with the ground, his body shattered. The other landed on a tree, unable to alter the steering in time; he incurred grave injuries during the fall. Now it would be my turn. I climbed on the edge of the balloon basket, ready to jump when ordered from below. I clearly spotted the enemy aircraft in a dive, approaching my balloon. The burst of fire from his MG narrowly passed above me, in between the balloon and the basket. Suddenly, a second MG rattled and the French aircraft dove directly toward me so that he almost rammed into me before crashing into the ground. A German Jagdeinsitzer saved me."[27]

The balloons were increasingly threatened by enemy aircraft which forced an increase in protection by MG and light anti-aircraft guns. Their own fighter aircraft did not offer sufficient protection, the German air force was hopelessly outnumbered in the battle of the Somme. Initially, Feldluftschiffer-Abteilungen utilized looted MGs; in addition to MG, each Ballonzug was equipped with a 3.7cm light anti-aircraft gun.

In early November 1916, battles on the Somme subsided, ultimately choking in the mud and blood of the soldiers. British losses counted 410,000, the French lost 200,000, and the Germans lost 500,000 men. By the end of August 2600 members of the German Air Force had fallen, 300 of which were Luftschiffer.[28]

Following the experiences of 1916, the Luftschiffer-Truppen were restructured. The battles of the Somme and Verdun revealed a lack of leadership and organization within the Luftschiffer-Truppen. In the future, two balloons would be assigned to each division's sector. At the suggestion of the Feldflugchef, Oberstleutnant Hermann Thomsen, the position of the Kommandierender General der

Luftstreitkräfte (Kogenluft) was created on October 8, 1916. On November 15, 1916, Generalleutnant Ernst von Hoeppner took over the position; Oberstleutnant Thomsen assumed the role of Chef des Generalstabes (Chief of General Staff). These moves allowed all Flieger, Luftschiffer, Flak, and the weather service (Wetterdienst) on the front and at home to form one agency.[29]

Endnotes

1 National Archives, Volume 13: *Die Tragödie von Verdun,* 1. Teil, Berlin 1926.
2 Reichskriegsministerium (War Office): *Die Operationen des Jahres 1916*, Berlin 1936.
3 Reichskriegsministerium: *Die Operationen des Jahres 1916*, Berlin 1936.
4 Hoffmann, Rudolf: *Briefe aus dem Weltkrieg*, München 1937.
5 Bavarian War Archives: *Die Bayern im Großen Kriege 1914-1918*, München 1923.
6 Erinnerungen von Hans Rödinger, Chemnitz. (VI. Reserve-Korps).
7 Vgl. *L'Aéronautique Militaire Francaise 1914–1918*, tome I, Paris 1980
8 National Archives: *Aufmarsch der deutschen Luftstreitkräfte zum Angriff auf Verdun*, Februar 1916.
9 Méchin, David: *The WWI French Aces Encyclopedia*, Vol. 4, Reno 2021.
10 Guttman, Jon: *Ballon-Busting Aces of WW1*, Oxford 2005.
11 Friedrich von Zanthier ruht heute auf dem Soldatenfriedhof Metz, Block 4, Grab 163.
12 Engelmann, Wilhelm: *Feldartillerie-Regiment Nr. 99*, Berlin 1927.
13 Wir Kämpfer im Weltkrieg: *Brief des Oblt. Linneborn, Feldartillerie-Regiment Nr. 99.*
14 Ludendorff, Erich: *Meine Kriegserinnerungen 1914 – 1918*, Berlin 1919.
15 On July 1, the German artillery counted 598 light and 246 heavy guns.
16 Fabeck v., Hans: *Im Orkan der Sommeschlacht*, Berlin 1930.
17 Probably a Fe2b of No. 25 Sqn. RFC.
18 Probably a Be2c of No. 2 Sqn. RFC
19 Vgl. *L'Aéronautique Militaire Francaise 1914–1918*, tome I, Paris 1980.
20 Vgl. Revell, Alex, *British Single-Seater Fighter Squadrons on the Western Front in World War I*, Atglen 2006.
21 To include Feldflieger-Abteilungen 1(b), 23, 32, and 59.
22 Royal Flying Corps Communique No. 40, June 1916
23 OHL: *Die Operationen des Jahres 1916*, Berlin 1936.
24 The 2nd Army had a total of 104 aircraft at its disposal.
25 *Das Ehrenbuch der Deutschen Feldartillerie*, Berlin o. J.
26 Ludwigshafener Frontkämpferbuch, Ludwigshafen 1936, S. 162.
27 On that day, Hauptmann Boelcke shot down the DH 2 (7895) of the No 32 Squadron near Thiepval.
28 Vgl. Reichsarchiv Bd. 20 & 21, Somme-Nord, Berlin 1927.
29 Neumann, Georg Paul: *Die deutschen Luftstreitkräfte im Weltkriege*, Berlin 1920.

Above: Heavy artillery, such as this 15 cm Krupp gun 16, was brought to the front by truck at Manancourt in the summer of 1916.

Right: Even the church of Bapaume was not spared during the fighting.

Below: Bapaume station in the summer of 1916.

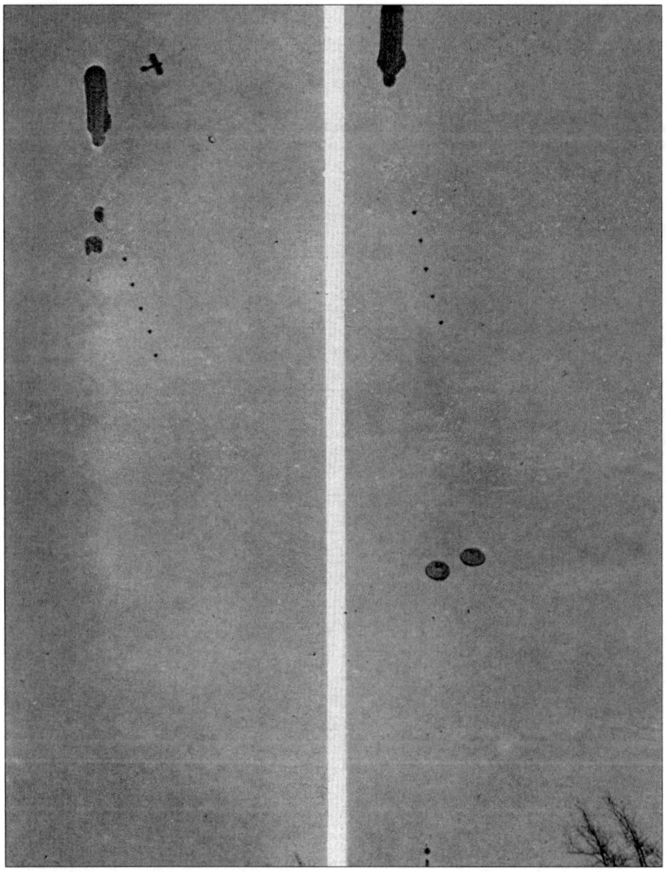

Above: Attack by an Albatros on a French balloon. The two observers jumped out with parachutes despite their low altitude.

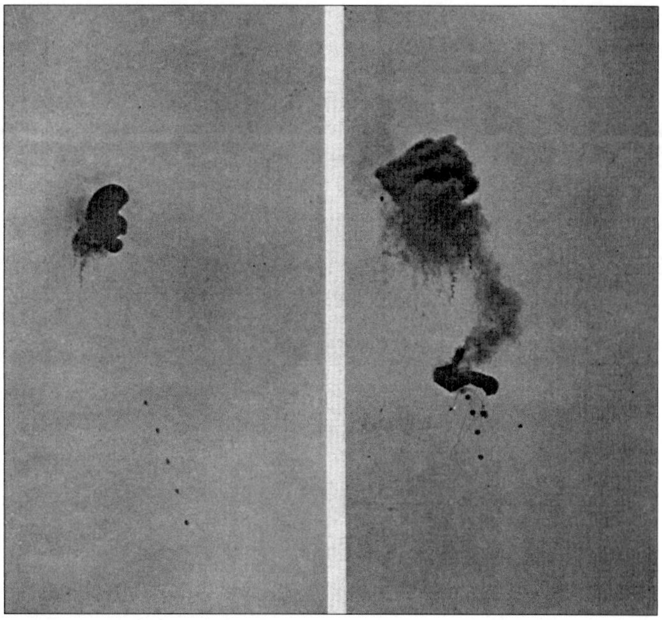

Above: Seconds later, the balloon is on fire.

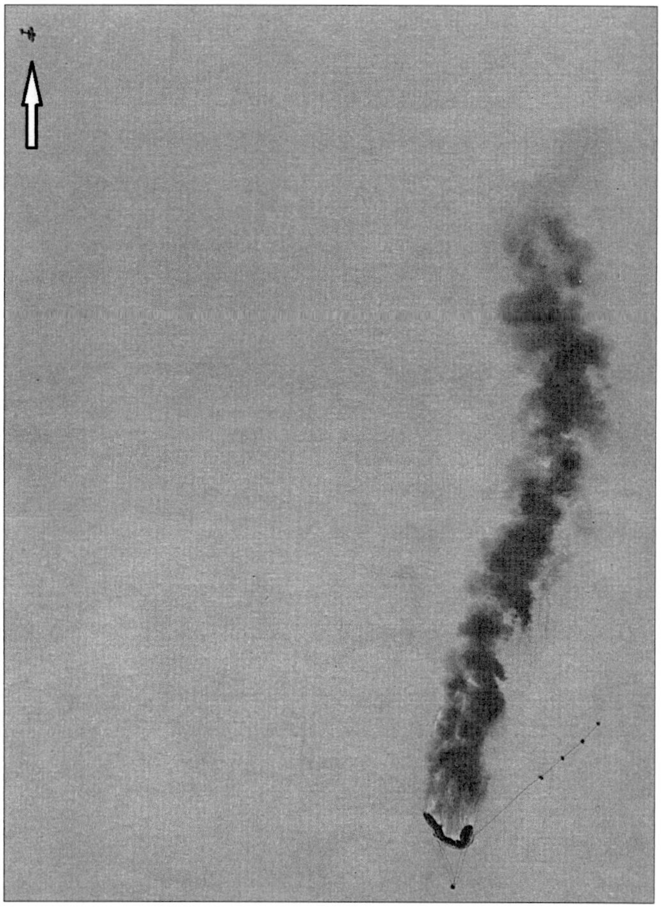

Right: Burning, the balloon falls to the ground, while the German Albatros flies back to its own lines.

The Year 1917

Above: During fall 1916, a British Caquot balloon drifted over the lines, which was immediately copied by the Germans as an Ae balloon.

Above: The Caquot balloon had several advantages, such as higher climbing performance and much smoother behavior in the air.

Above: The Parseval-Sigsfeld type still had simple sails on the rear hull as control surfaces.

Above: Group photograph of Bz 2 with the 4th Army in Flanders with a Parseval-Sigsfeld balloon.

Following the end of the battle of the Somme, the Ballontruppen were equipped with the Ae-balloon, which was the next technological advance in balloon development. In the fall of 1916, the British RFC Kite-Balloon-Service had captured a Caquot balloon; instead of control surfaces, this type of balloon had three air filled steering bulge cores, allowing it to be almost horizontal when airborne. It reached a height of more than 1600 meters. The German as well as Austrian-Hungarian Feldluftschiffer-Truppen utilized the balloon under its military designation type, Ae (A= 800, e=English). With the introduction of the Ae balloon followed the utilization of HP driven motorized winches (80-100HP) as well as parachutes which left the Ballontruppen well equipped for the upcoming battles of 1917. In addition, anti-aircraft artillery was improved with by adding more MG and 3.7cm Krupp anti-aircraft machine guns.

In the course of the re-organization of Kogenluft Generalleutnant Ernst von Hoeppner, the previous Feldluftschiffer-Abteilungen were restructured into 53 staff departments, 128 Ballonzüge[1], and seven balloon headquarters from March 26, 1917. Ballonzüge deployed with only one balloon each, allowing them to stay mobile and be transferred quickly to a different front sector. It was subordinate to one staff department which in turn was attached to a balloon headquarters. The staff departments stayed on the ground, close to the Army; they were able to request Ballonzüge when necessary.

Luftschiffer (Stoluft) staff officers organized a reporting system within balloon headquarters. A multitude of reports were received from the Army balloons; they were sorted according to their urgency. Important messages were

Above: The balloon of Bz 2 is filled with gas.

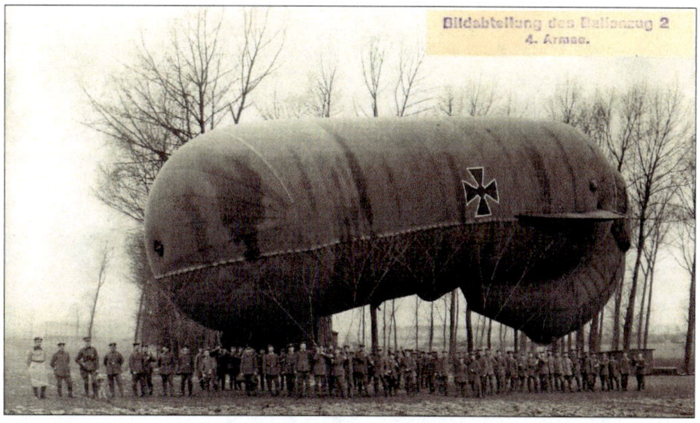

Above: However, the Parseval-Sigsfeld balloons remained in use until the end of the war.

Above: An MG 08 in a rotating mount for defense against enemy aircraft. (NARA)

Above: This double gun mounting for two MG 08s was produced by the field airmen themselves. (NARA).

immediately passed to AOK, artillery, or infantry. The Ballonzüge in turn received weather service information from balloon headquarters, including temperature, pressure, or thunderstorm forecasts.[2] Stoluft Order Command included a Feldluftschiffer-Park which was in charge of handling Ballonzug supplies. Before the implementation of this department, supplies had to be requested with the respective Luftschiffer-Truppe (Iluft) Inspektionen. Stoluft reported the observations of their balloons to AOK every day; Stoluft in turn evaluated the reports and passed them on to troops at the front. Below is a list of reports from June 1917:

June 23: 4th Army- 125 old batteries
6th Army- 133 old and eight new batteries
7th Army- 170 old and six new batteries
1st Army- 74 old and three new batteries

June 24: 4th Army- 124 old batteries
6th Army- 247 old and four new batteries
7th Army- 128 old and seven new batteries

The 6th Army reported an increase in enemy artillery and heavy railway traffic was observed in both directions behind the frontlines, indicating that an attack was imminent.

Bavaria's newly established Ballonzüge b 5 and b 6 were to be deployed to the eastern front and therefore were added the initial **O** (O for Ost= east, i.e. b 5 O). Balloons deployed to the western front were assigned the initial **W**.[3] The only difference between Ballonzüge **O** and **W** was the difference in personnel numbers. One Ballonzug-West counted eight Offiziere, two Unteroffiziere, and 69 enlisted. One Abteilungsstab West consisted of 12 Offiziere and 32 enlisted, including doctors and medics. As Bavarian

Abteilungsstäbe and Ballonzüge were distinguished from the Prussian ones by the prefaced **b** in written correspondence, misconceptions were common.

As a result, a correction was made to the designation of the Bavarian units on June 16, 1917. Department staffs were changed from 1-8 to 61-68, and Ballonzüge 1-23 to 201-223.

On July 30, 1917 Kogenluft ordered yet another restructuring of the Luftschiffer-Truppen at the front:[4]

- Stoluft positions will be dissolved.
- Each army will be assigned a Kommandeur der Luftschiffer (Koluft) and one Ballonzentrale.
- Feldluftschiffer-Park 5 in Valenciennes is subordinate to Koluft 5 at AOK 6.
- Feldluftschiffer-Park 1 in Sedan is subordinate to Koluft at AOK 1.
- Feldluftschiffer-Park 2 in Lahr is subordinate to Koluft

Above: The entire staff of Bavarian balloon train 208.

Above: The crews and sergeants of Bavarian Balloon Platoon 208 according to the new numbering, which was introduced in the Bavarian units on June 16, 1917.

Above: The camouflaged tents and barracks of Bz 208 on the Aisne; the balloon has risen some distance away.

Above: The gas teams of Bz 208 were only allowed to smoke at a safe distance from the gas cylinders.

Above: With the Blinkgerät 17, Lt. Wagenscheid was able to transmit orders to the infantry.

AA C.

In addition to his duties as the commander of FLA 65, Hauptmann Koben was assigned the position of Koluft at Armee-Abteilung (AA) C. Ballonzüge deployed for training at the firing ranges along with the artillery were subordinate to the Koluft during that time.

Koluft 6 – Schießplatz (firing range) Thimougies
Koluft 2 – Schießplatz Maubeuge, Jurbies, Sebourg
Koluft 1 – Schießplatz Signy l'Abbaye, Maubert-Fontaine
Koluft 5 – Schießplatz Longuyon
Armee-Abteilung A – Schießplatz Insmingen

Kogenluft was particular about further training. In order to create practical training opportunities, the Feldluftschiffer-Schule (training center) was transferred from Jüterbog to Namur in Belgium. The threeBallonzüge assigned to the Luftschiffer-Lehrabteilung were issued designation numbers 159, 160, and 161 on December 27, 1917.

Work at balloon headquarters was also in need of simplification. During active battles, a multitude of incoming messages and reports had to be processed and sorted; as a result, balloon registration stations were established within individual Armeegruppen (General-Kommandos); here, messages sent by observers were filtered according to their importance and passed on to the respective balloon headquarters.

Infantry balloons were tasked with the support of fighting ground troops, specifically for the infantry division. This decision was made as in times of thick ground fog or low clouds, light signals ordering the discontinuance of their own artillery fire were often difficult if not impossible to be

Above: Flashing device with battery, this was omitted after Lt. d.R. Hampel had constructed a dynamo on the balloon winch to generate electricity.

Above: One observer has landed safely, still wearing the chest phone and earphones.

detected by storming infantry soldiers. Balloon observers in turn detected the light signals more easily from their elevated positions and immediately passed them on to the battle registration offices. Infantry also used the Zeiss- Blinkgerät, M-Blink 16, or Blinkgerät 17 (signaling equipment) to

Above: Infantry flasher unit with an M-Blink 16 flasher.

Above: On July 22, 1917, the balloon of Blinkerzug 46 crashed west of Pleine-Selve.

Above: The field railroads (60 cm track gauge) were laid by the pioneers of the railroad troops along the entire front. This allowed troops and material to be transported right up to the front line.

transmit messages to the balloons or receive messages from balloon observers.

From early July, 1917, Blinkerzug 46 was under the command of Uffz. Heimann; they were stationed west of Pleine-Selve on the Aisne river. At this location, balloon signal messages were received and transmitted via telephone

Above: The light railroads had direct connections to the stations, as here in Manre in Champagne.

to the Saxon Fernsprech-Abteilung 208. On July 22, 1917, Gefreiter Franz Müller was taking a break from his duty at the M-Blink 16-Gerät[5]. Suddenly, he heard shots being fired by the Ballonzug's 3.7cm anti-aircraft artillery, aiming at a French aircraft squadron:[6]

"One aircraft separated from the squadron and dove toward our tethered balloon. Hastily, the men tried to lower the balloon, but the aircraft was faster, passing overhead at 600 meters. The flammable ammunition ignited the balloon; the observer managed

Above: On the army training grounds, the personnel of the field airship divisions regularly completed training in new tactics. This observer uses the peaceful time in the basket for gymnastic exercises.

to jump out of the basket with his parachute just in time. The pilot was not satisfied with setting the balloon on fire, he turned the aircraft around and dove toward the parachuting observer at 300 meters. Despite heavy gun fire, the aircraft managed to escape. The

Above: A carrier pigeon cage was also attached to this balloon basket. (NARA)

balloon, sank to the ground in burning shreds, the observer slowly glided to the ground wearing his parachute, heavily injured by multiple MG hits."

The French Nieuport were part of Escadrille N 78, Slt. Pierre Le Roy de Boiseaumarié was credited with the aerial victory.

In December of 1917, Lt. d. R. Hampel of the FLA 25 constructed an improved signaling device using metal mirrors. Hampel attached dynamo-powered generator to the winch which provided the lamp with a steady supply of electricity, omitting the use of batteries altogether. Following extensive experimentation which Hampel carried out with Bz 25, he submitted a suggestion for improvement directly to Kogenluft which was subsequently implemented industry wide.[7] Lt. d. R. Hampel also designed special light railway transport cars (track width of 60cm) for the Ballonzug. The light field railway system was established by the pioneers of the Eisenbahntruppen; it was capable of withstanding even the most adverse weather conditions. Hampel constructed railway cars for the transportation of winches, gas bottles, balloon casings, anti-aircraft guns and mobile phone networks, as well as signaling devices on tripod stands. This invention allowed for the rapid relocation of a Ballonzug while avoiding the road system.

The Prussian FLA 7 and Bavarian FLA 207 submitted a suggestion with Kogenluft. In the past, their balloons had not been lowered during an advance but rather they were pulled ahead (Ballon hoch) above the roads, covering distances of up to 8km at times to get closer to the frontlines. The multitude of telephone lines erected alongside or across the roads proved to be a challenge to these advances. For the balloons to more easily advance towards the sites of upcoming battles, telephone lines were no longer allowed to be established close to any roads from early January 1918. The distance between the road and any phone lines had to measure at least 50 meters. If plans were made for a phone line to cross a road, it had to be established underground. This allowed for the unobstructed relocation of balloons. As it turned out, many units did not heed this ban in practice.

On April 6, 1917, US President Wilson declared martial law on Germany; on April 10, a war declaration on Austria-Hungary followed. On April 20, the war with Turkey began. A new opponent had entered the stage of war, one who would ultimately decide the fate of WWI. From April 9, 1917, the British offensive near Arras was stopped, despite heavy losses. The French attacks on the Chemin de Dames (also known as "The Sanatorium of the Western Front") and the Champagne region from April 16 were unable to achieve their desired breakthroughs. The first step of the attack was for artillerists to eliminate the German balloons. In return, balloon observers attached particular value to combatting guns specifically designed to fight balloons. These guns aimed directly at balloons, shooting fuse grenades so that the 5th through 7th round hit the balloon. As soon as one of these batteries opened fire on a balloon, observers in turn directed their own artillery to open fire on the attackers. Simultaneously, ground troops pulled the balloons sideways across castors, allowing the balloons to avoid enemy gun attacks both vertically and laterally. On August 19, 1917, artillerists of the 5th army bombarded French balloons using burning fuses and percussion fuses. Percussion fuse grenades yielded better success than the usual burning fuses, particularly after the balloons had been lowered closer to the ground.[8]

On June 7, Marschall Haig ordered enormous mine blasting to begin the Battle of Ypres (Flandernschlacht). The Wytschaete ridge was lost, and, on July 31, attacks were expanded to include the areas of Ypern through Dixmude. The Battle of Ypres did not end until November; attacks subsided, with only a few kilometers of territorial gain towards the East. The situation on the western front appeared to calm down, according to a report from a 5th army Koluft on the Verdun front on November 10:

„Enemy balloons showed a remarkable inactivity at that time. Despite good visibility, only half of the available (balloons) had ascended. [...] The French are typically completely inactive in bad weather conditions."

OHL concluded that, there would not be any further enemy attacks in the vicinity of Verdun for the foreseeable future. The 2nd army under the command of General v. der Marwitz was stationed near the road section from Cambrai-Arras to La Fère. The quiet part of the front led them to believe that no enemy attacks were imminent. While a German infantry patrol across enemy lines on Sunday, November 18 resulted in captives from several British divisions who told of a planned attack, with winter approaching, all seemed quiet on the "calm front near Cambrai". The surprise was all the greater when, on November 20 at 07:15, the British artillery opened the Battle of Cambrai (Tankschlacht von Cambrai) using enormous artillery fire. While artillerists were still firing, British tanks began to move, rolling across German lines. They did not take advantage of their advance to Bourlon-Wood, Noyelles, and Honnecourt; by December, they had been pushed back to Graincourt-Marcoing-Honnecourt.

German artillery supported FLA 34 using Bz 201 and 202, FLA 8 using Bz 28, 30, and 209. They supported FLA 2 with Bz 2, 50, 56, and 137, as well as FLA 5 with Bz 14 and 210.

As soon as the British artillery fire started, the Bz 50 balloon ascended northeast of Rumilly; however, it soon had to be pulled down again due to continuous air attacks. As the British infantry approached, Ballonzugführer Lt. d. R. Kerstenhann ordered his men to retreat to Niergnies. At 16:00, British tanks were stationed at Rumilly, allowing only the balloon to be saved. Vehicles and gas bottles had to be left behind. The German infantry was unable to enter Rumilly again until several days after the initial attack; Kerstenhann managed to salvage part of the equipment and materials.

Above: Retrieving the balloon after the exercise, the horse-drawn carriages in the foreground suggest a peaceful environment.

Facing Page, Middle: Lt. d.R. Kerstenhann led Bz 50 in the tank battle of Cambrai.

Left: Photographing with the 120 cm camera in the narrow basket required a lot of practice.

Above: Group photo of Bz 50 in Rumilly in October 1917.

Before November 20, 1917, the front near Cambrai seemed to be a quiet place, Lt. d.R. Kerstenhann relaxes with a pipe in his quarters.

Above: On July 24, 1916, this German balloon drifted into the French lines.

Ae balloon near Cambrai in October 1917.

Above: A German balloon in front of the Piave Valley. The ascent in the mountains placed particular demands on man and material. (NARA)

Above: Italian aircraft, such as the Caproni Ca. 3, bombed the German and K.u.K. troops even before the battle. This Caproni was shot down by a German Jasta.

Balloons in the High Mountains of the Italian Front

In the summer of 1917, Austro-Hungarian K.u.K. troops showed initial signs of weakening. The battles of the Isonzo and the 1916 Tirol Offensive were aborted following some heavy losses. After the death of Emperor Franz Joseph on November 21, 1916, the Danube Monarchy, specifically the Czech Bohemia showed aspirations for independence.

To assist the Allied Forces, the decision was made in September of 1917 for OHL to support K.u.K. troops in their attacks north of Isonzo (12th battle of Isonzo).[9] For this purpose, 2400 trains transported materials including 68,000 horses to the front at Isonzo.[10] Erzherzog Eugen von Österreich was the chief commander of the Southwestern Front. He was in command of the K.u.K., 10th army, the 1st and 2nd Isonzo armies, as well as the 14th German army. The newly established 14th army consisted of six infantry and fighter divisions including 540 guns and field artillery, 16 Flieger-Einheiten and Ballontruppen. Ballonzüge 84, 85, 87, 103, 113, and 114 were subordinate to the German Feldluftschiffer-Stäbe 33 and 45.

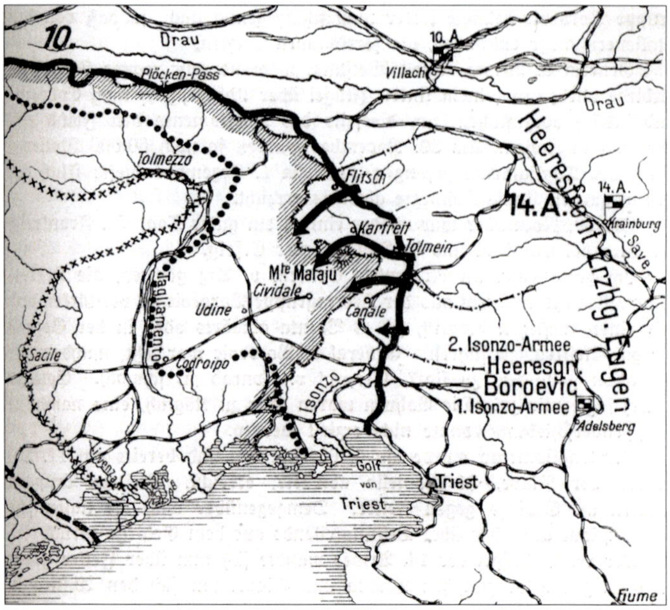

Above: Map of the offensive between Flitsch and Canale in October 1917.

A Feldluftschifferpark commanded by a Koluft (Kommandeur der Luftschiffer) was charged with supply replenishment. The high mountains including Mount Matajur (elevation 1643 meters) posed completely new challenges for the Feldluftschiffer. As a result, the Ballonzüge were deployed to the Italian front in mid-September to practice mobile warfare in mountainous regions. Balloons of the first group of attack had to set up ascent points at an elevation of 1000 meters to reach ascent heights of 2000 meters. This in turn meant that all the equipment, materials, balloons, winches, and gas cylinders had to be transported to the ascent locations via two narrow trails. Two Ballonzüge each shared one ascent location.

Until October 1, AOK 14 forbid German aircraft from overflying the front on grounds of secrecy. Thereafter, the Italian area was opened for air reconnaissance carried out by Flieger-Abteilung 14 and Reihenbildzug 1. Italian leadership was aware of the attack preparations; Czech and Romanian officers had deserted and shared details regarding the upcoming attack. An increase in aerial reconnaissance was a telling indication of the planned offensive. The bomb attacks that followed were obstructed but not completely prevented. The Austro-Hungarian Fliegerkompanien-Jagd (Flik J) supported the German Jagdstaffeln 1, 31, and 39 in strengthening their defense. Rumors spread quickly among the ground troops that it was Jagdgeschwader Richthofen, due to the remarkable red color of their aircraft fuselage.[11] However, it turned out to be Hptm. Godwin Brumowski (Flik 41J) who flew an Albatros with a red fuselage.

The Flitscher group of the K.u.K. army, commanded by General Krauss, and the 14th German army under the command of General Otto von Below were to advance toward Udine between Flitsch and Canale. On a windy,

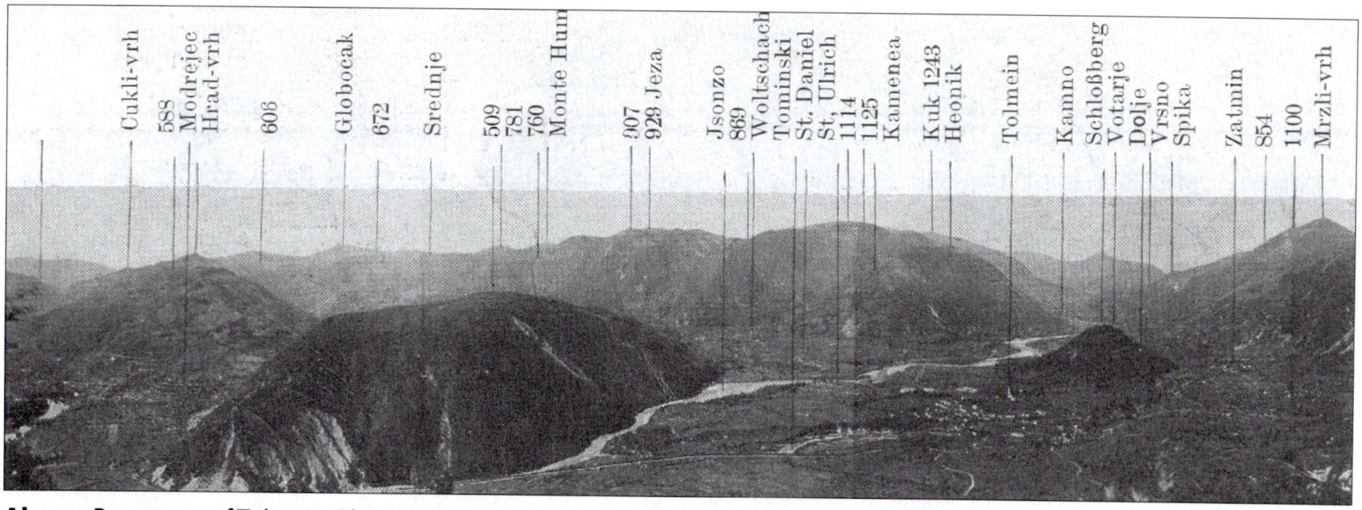

Above: Panorama of Tolmein. The picture gives an idea of the problems faced by the balloon trains.

snowy October 24, 1917, the first day of attack, the balloons were utilized as follows:

Ballonzug 84 - Gruppe Scotti (K.u.K. XV. Korps)
Ballonzug 87 – Gruppe Berrer
Ballonzug 102 – Deutsches Alpenkorps (German Alpine Corps)
Ballonzug 103 – 12th Infantry Division
Ballonzug 113 – AOK 14 Staff
Ballonzug 114 – Gruppe Stein (III. bayer. AK)

As the snowstorm on October 24 prevented any observations from the ground, Staff and artillery relied exclusively on reports from the balloons. Thanks to the extensive preparation work such as setting up telephone lines, balloon observer reports arrived nearly without any time delays.

After the crossing of the Isonzo river, the reserve balloons were utilized directly on the front lines while the attack group's balloons remained at their respective ascent points, leading the artillery of the 14th army's southern flank. Ballonzug 87 (FLA 45) led the K.u.K. army's artillery toward the road to Isonzo near Canale, which allowed the artillerists to advance quickly. The 2nd Italian army barely had any time to retreat. 60,000 Italians surrendered near Cadroipo on the river Tagliamento on October 30.

There were delays during the advance towards Udine as all troops using heavy machinery and equipment was forced to use the only road via Karfreit passing through the Natisone valley at Cividale. Only Ballonzug 103 (FLA 45) was able to reach Tagliamento in time; there, the Italian troops had been surrounded by German troops east of the river.

At the time of the crossing of the river Tagliamento, all balloons were available again; however, not all of them needed to be utilized as there were hardly any battles fought at that point. After the Piave was reached, the first French and British aircraft began attacking the balloons. Aircraft support had been hastily requested to assist the Italian allies. The first losses occurred due to continuous attacks on the German balloons, in spite resistance by MG and 3.7cm automatic cannons. Equipment and other supply replenishment proved to be a challenge as the location of the Feldluftschifferpark was 250 km away. For the time being, supplies were catered via truck, which was a time-consuming task. Later, the railway was used to transport more supplies via Pontebba- Gemona- Pardone.[12]

Mid December 1917, the offensive on the shores of the Piave stopped and the 14th army returned to the western front.[13] The German and Austro-Hungarian losses totaled 70,000 men; Italians lost approximately 340, 000 men as well as large quantities of war material.

Endnotes

1. According to the Army Research Institute for War History, 150 Ballonzüge were stationed at the front at the end of 1917, utilizing one balloon each. Vgl.: Oberkommando des Heeres, 12. Band, Berlin 1939.
2. Bavarian State Archives, Department IV War Archives: File ILuft 179.
3. Bavarian State Archives, Department IV War Archives: File MKR 1364.
4. Bavarian State Archives, Department IV War Archives: File ILuft 180.
5. Indicated as letter G (Graben, trench) Sig-16.
6. Müller, Franz: Mit dem Blinkerzug Nr. 46 an der Westfront, Kiel 1932.
7. Bavarian State Archives, Department IV War Archives: File ILuft 41.
8. Luftstreitkräfte newsletter Nr. 28, September 1917.
9. Ludendorff, Erich: Meine Kriegserinnerungen 1914-1918, Berlin 1919.
10. Alt-Kaiserjägerclub: Vom Isonzo zur Piave, Innsbruck 2007.
11. National Archives: Der Durchbruch am Isonzo, Teil I, Berlin 1926, S. 99.
12. National Archives: Der Durchbruch am Isonzo, Teil I, Berlin 1926.
13. Formerly referred to as 17. Armee.

Before October 1, 1917, German aircraft were not allowed to fly over the Italian front. Until then, the Italian bombers mostly fought the flak, like this 7.7 cm Flak L/27 on Krupp trucks. (NARA)

Above: All the supplies for the 12th Battle of the Isonzo had to be transported by truck over narrow mountain roads. (NARA)

Right: 120 cm searchlights were used to illuminate the front. Enemy positions and batteries could also be spotted from the mountains at night using searchlights. (NARA)

Below: Oblt. Frank Linke-Crawford flew Oeffag Albatros 153.16 from October to December 1917 with Fliegerkompanien-Jagd 41 (Flik J) on the Isonzo Front.

Oeffag Albatros 153.16 after delivery to the Luftfahr-Truppe in August 1917.

Above: Blown up Caproni Ca. 3 found by German and Austrian troops at the Campo-Formido airfield.

Above: Commemorative photo of Balloon Zug 84 in Veneto on January 4, 1918 before the transfer to the Western Front.

1918, The Last Year of the War

Above: In the spring of 1918, Lt. Wagenscheid of FLA 208 was promoted to an auxiliary observer. After training at the Namur Field Airship School, he was deployed to the front under the guidance of an experienced observer.

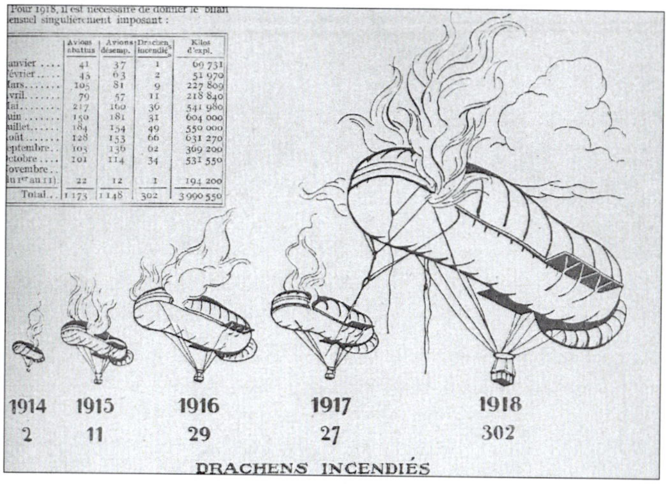

Pour 1918, il est nécessaire de donner le bilan
ensuel singulièrement imposant :

	Avions abattus	Avions désemp.	Drachen incendiés	Kilos d'expl.
anvier	41	37	1	69 731
évrier	45	63	2	51 970
Mars.......	105	81	9	227 809
Avril.......	79	57	11	218 840
Mai........	217	160	36	541 980
uin	150	181	31	604 000
uillet......	184	154	49	550 000
oût	128	153	66	631 270
eptembre..	103	136	62	369 200
ctobre	101	114	34	531 550
ovembre .. (du 1er au 11)	22	12	1	194 200
Total...	1 173	1 148	302	3 990 550

Above: In 1918, 315 German balloons fell victim to enemy aircraft. According to the list from 1919, 302 of these were French airmen. However, Belgian ace Willy Coppens claimed 35 balloons in his 37 victories and American ace Frank Luke claimed 14 balloons among his 18 victoies.

In the winter of 1917/1918, the Kogenluft issued new tactical orders for the balloon troops for the year 1918. Due to the improved equipment and training of personnel, not all balloons of each field airship division needed to be launched simultaneously in the coming battles. A few balloons were sufficient, as the attack fronts also became much narrower. One balloon was enough for two divisions. The reserves formed were then used during the attack at the firing points or as replacements for balloons that had been shot down. As a precaution, the balloon platoons were given a second balloon and gas cylinders that lasted for a week. In addition, the mobile gas stations on railroad cars ensured a constant supply.

In the spring of 1918, Major Gundel, as head of the Airship Troop Inspection (Idluft), was assigned fresh personnel from the Army Airship Corps. The army had handed over the Zeppelin and Schütte-Lanz airships to the navy and disbanded the troops. As a result, experienced technical personnel supplemented the field airship detachments and balloon platoons. In the airship replacement sections (LEA), the balloon master workshops were established, in which the damaged balloon envelopes and baskets were expertly repaired by trained sailmakers. However, there was an increasing shortage of rubberized materials, which had to be replaced by synthetic rubber. The six replacement departments continued to train new personnel, so that by the summer of 1918 the balloon platoons had increased to 186 and the department staffs to 56. The losses of the field airmen increased rapidly in 1918. Of the total of 471 balloons shot down, 315 were lost in 1918 alone. 132 balloons were burnt in September 1918 in aircraft attacks and 37 balloons were destroyed by enemy artillery.

Above: Industrial tractor with balloon winch and telephone cable reel, so that the telephone connection to the balloon reporting centers could be maintained even on the march.

To enable the balloons to follow the storm troops in the offensives of 1918, even in difficult terrain with funnels and trenches, a replacement had to be found for the horse-drawn carriages. Industrial tractors, for which rough terrain was no obstacle, were used for this purpose. This allowed the high-altitude balloons to follow the infantry slowly and continue

Above & Below: In June 1918, a balloon is prepared for transportation with a Daimler tug near Soissons. (NARA).

Above: Anchored, the balloon rises from the Daimler tractor. The balloon can be moved easily in the open terrain. (NARA).

Above: The airmen's trucks could only move the balloons in high transport over paved roads or flat terrain. (NARA).

to observe them. On March 8, 1918, the Bavarian Balloon Force Train (BKz) 206 moved to the front. This balloon train, which was equipped exclusively with trucks, motor vehicles and motorcycles, was the first to completely dispense with horses. From March 11 to 27, the BKz 206 made frequent ascents in the area north of Reims in order to conceal the Michael Offensive, also known as the Kaiserschlacht, between Arras and Le Fère. The 8500 kg Daimler power winch was also able to overcome difficult terrain thanks to its special iron wheels (gripper system). The Telephon truck also had these iron wheels and thus the connection to the balloon control center was secured. The leader of the BKz 206, Lt. d.R. Deuschle, only complained about the late participation in the Michael Offensive from March 30, 1918. In April 1918, Koluft 18 transferred the BKz 206 to the 231st Infantry Division.

In the area of the 4th Army in Flanders, the balloon platoons also increased the number of reserve balloons from the beginning of March to disguise the Michael Offensive. Here, the increased number of visible balloons was intended to simulate an imminent attack. In many baskets, however, there were only straw dolls instead of an observer. However, as there was only one winch for each balloon train, the balloons were anchored to ground anchors. When a Bz 2 balloon broke loose and drifted towards the British lines, there was great excitement and the anti-aircraft and artillery tried to shoot it down. In the end, the British artillery destroyed the balloon in no man's land.

The German Offensive in March 1918

The decisive battle of 1918, the Michael Offensive, extended 75 km between Arras and Le Fère. The 17th and 2nd Armies led the main thrust, while the southern 18th Army accompanied the attack as far as the Somme and then took over the flank protection. Five balloon units with 15 balloon platoons were available, 11 of which were assigned to the attack troops. The remaining 4 were grouped into 2 marching platoons, which were to be deployed at the focal points during the course of the battle. Artillery fire was of crucial importance for the advance of the infantry. For every kilometer of attack front, 20–30 batteries, approx. 100 guns, were deployed. However, large-scale artillery fire had to be avoided in order not to give away the attack intentions. For this reason, only individual guns fired at their subsequent targets with the help of the balloon observers. The King Charles 1st Württemberg Field Artillery Regiment No. 13 took up position in the attack area of the 27th Infantry

Above: Balloon picture of Reims in March 1918, taken from a distance of 10 km. The cathedral is in the center of the picture.

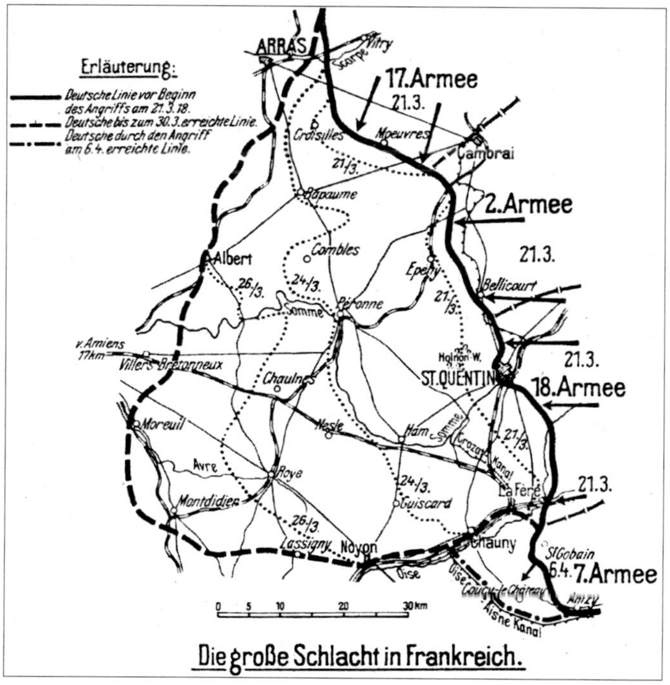

Above: Map of the Michael Offensive, also known as the Imperial Battle.

Above: Private Johannes Gronbach took part in the Michael Offensive in March 1918 as an artilleryman in the King Charles 1st Württemberg Field Artillery Regiment No. 13.

Division south of Villers-Guislain from March 10. Only one gun from each battery of field guns was brought into the position during the night. Private Johannes Gronbach wrote about this:

"Flieger-Abteilung 227 provided us with aerial photographs of the attack area and our captive balloon, which stood alone in the sky, shot down the entire artillery of the 27th Infantry Division. Only one shot was allowed to be fired, which was agreed in advance with the observer in the balloon. The observer then informed us of the firing position by telephone."

When the assault began, the artillery was to lay down its fire in front of the infantry and lead the way as a so-called "fire roller". The shorter the distance between the fire roller and the infantry, the less time the enemy had to leave their shelters. However, the precisely calculated time schedule of the artillery often did not coincide with the advance of the infantry during the battle. Either the infantry needed more time due to the heavily battered terrain – 1 hour was calculated per kilometer – or they were caught in their own artillery fire due to the faster advance. Here the balloon observers had to keep a close eye on the current front situation and immediately report back to the balloon reporting points. Despite the most precise preparation, the infantry later suffered repeated losses from their own artillery when they ran away from the fire roll. The position signals emitted by the flare guns were probably overlooked in the smoke and fumes of battle.

On March 21, while it was still dark, the balloons rose about 3 - 5 km behind the front line. The artillery fire began at 4.40 a.m. and at 9.40 a.m. the infantry emerged from the trenches. The balloons followed immediately and continuously reported the positions of the enemy troops, although the battlefield was partly covered by dense ground fog. The airships often proved to be decisive in

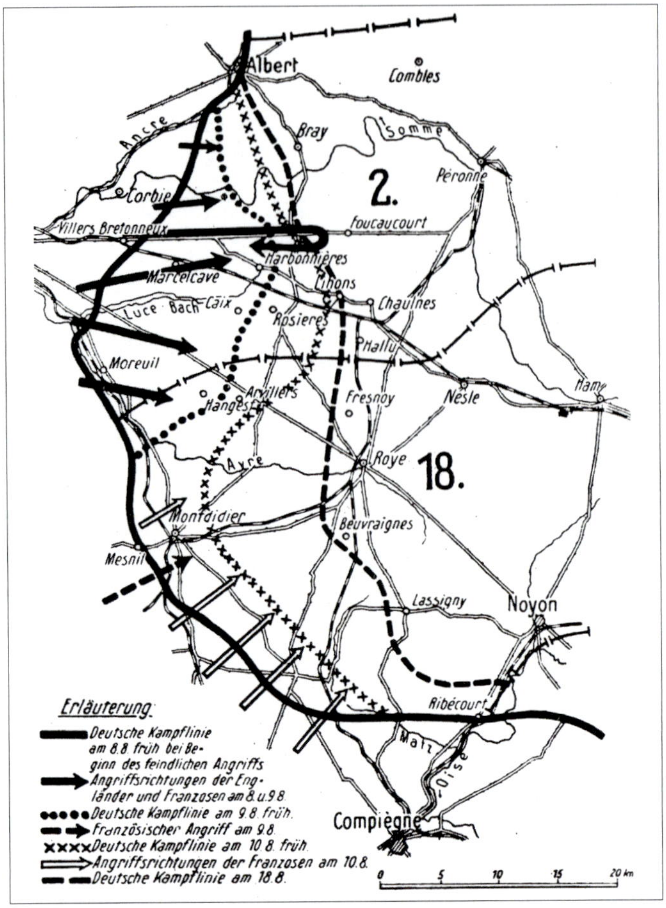

Above: On August 8, 1918, French and British troops unexpectedly broke through the German front between Albert and Ribecourt.

Above: The German balloons of the 2nd and 18th Army suffered heavy losses on the first day of the battle, and only rarely were they able to ascend without being immediately attacked by Allied aircraft.

the Michael Offensive, especially the Bz 14 and 56 of the FLA 2 delivered important reports to the command in a timely manner. The balloons followed the infantry in high transport, despite enemy fire. The practical training provided by the airship school in Namur, which had trained not only observers but also section and platoon commanders, now proved its worth.

Bz 96 had the winch pulled by 10 horses to transport the balloons over the soggy ground. 38 power and telephone lines that had been laid across the roads proved to be a particular hindrance. The balloons also crossed the town of St. Quentin on March 23rd in the high transport, whereby 51 lines had to be overcome.

Over the next few days, the balloons followed the infantry closely and also crossed the Scheldt. On March 24, the balloons of FLA 2 continuously reported the front line, which had been marked by the infantry with flashing signals and white cloths laid out. Thanks to the observations of Bz 209 (FLA 15), enemy batteries could be engaged directly and the marching train of Bz 72 and 151 advanced as far as Péronne. On March 25, the balloon platoons of FLA 10 finally crossed the Somme together with the infantry. Until the end of the month, the balloons always followed the infantry and thus provided valuable reports, although storms and rain made the observers' work more difficult. Near the front, enemy fire on the balloons increased sharply and the fighter planes also attacked more frequently. At the beginning of April, the German offensive came to a standstill and it was possible to penetrate up to 60 km into the British and French positions over a width of 75 km. But the hope of a decisive breakthrough was not fulfilled and soon the troops were back in the trenches facing each other.

The Catastrophe of August 8, 1918

On August 8, 1918, the French 1ème and 3ème Armies, as well as the British 4th Army with Australian, Canadian and American units, opened the offensive against the German 2nd and 18th Armies in the area between Albert and Montdidier. At Villers-Bretouneux, Canadian, American and Australian divisions penetrated deep into the German positions. The German 2nd Army in particular was practically overrun. It was not until 10 days later that the front could be consolidated again on the Harbonniers and Ribécourt line.

In the days leading up to the offensive, dense fog and heavy rain prevented German aircraft from conducting aerial reconnaissance. As a result, Flieger-Abteilungen 3, 28, 40, 224, 219 and A 269 only rarely monitored the enemy area between Albert and Montdidier until August 7. The FA 33 and 217 were even withdrawn from the 2nd Army area on August 4 due to the calm front situation and moved to the hinterland to recover. The German infantrymen in the trenches regularly heard sounds of enemy attack preparations during the night and reported this to AOK 2 and 18. During the day, however, the balloons were unable to detect anything conspicuous. On the one hand, the poor visibility prevented observation and, on the other, the Allied camouflage worked perfectly.

On the evening of August 7, 1918, the airmen of AOK 2

Above: A German balloon falls to the ground on fire.

Above: The balloons of the RAF and the Compagnie d'Aérostiers received constant protection from their own aircraft during the battle. The motorization by the Delahaye winches also allowed the balloons the greatest mobility.

Above: Given the Allied superiority, the balloons of the Compagnie d'Aérostiers had little to fear from German fighter planes.

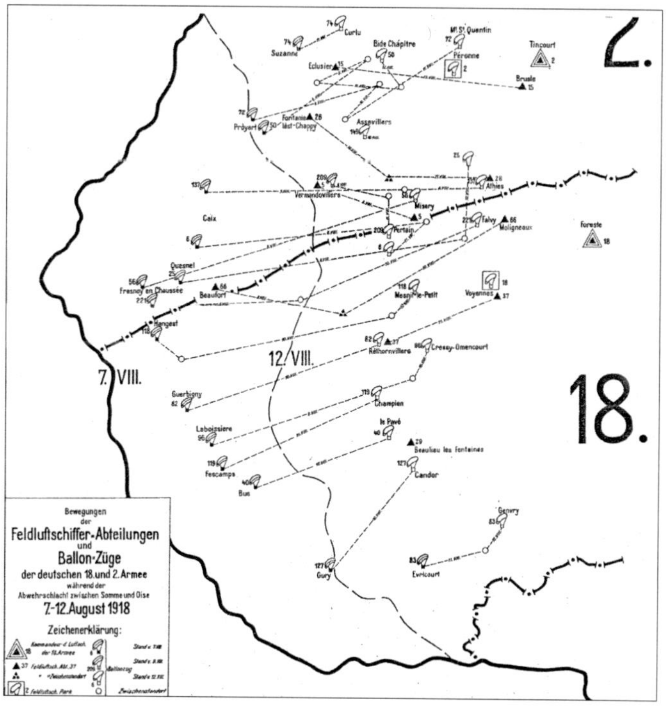

Above: Distribution of the German balloon troops in the 2nd and 18th Army in August 1918.

consisted of:

- Koluft 2 Major Gaißert with balloon center and field airship park in Foreste
- FLA 2 (Rittmeister v. Hatten) with Bz 74 and 151
- FLA 15 (Rittmeister v. Rochow) with Bz 50, 72 and 149
- FLA 26 (Major Schoof) with Bz 6 and 137
- FLA 5 (Hptm. Haeubert) with Bz 25, 56, 209 and 221

The platoons each had a 3.7 cm Maschinen-Flak platoon at their disposal. A further 93 anti-aircraft guns in 3.7 cm - 8.8 cm caliber were distributed behind the front of AOK 2. The 18th Army only had 49 guns in 7.7 cm - 8.8 cm caliber.

The British 4th Army had the No. 5 Balloon Wing with the 13th, 14th, 15th and 16th Companies. The French 1st Army had the 30th, 31st, 38th, 63rd and 81st Compagnie d'Aérostiers. All balloon units were motorized and therefore well prepared for the coming war of movement. On the evening of August 7, the pilots of the Escadrille de Chasse received another explicit order from the Commanding General of the 1ème Armee, Marie Eugène Debeney, to destroy the German balloons at all costs.

On the morning of August 8, the Allies launched their attack across a front of 35 km. Artillery fire began at 4.30 a.m., followed immediately by tanks and infantry. The British 4th Army advanced 18 kilometers and the French 1ème and 3ème armies 10 kilometers. The Germans were completely surprised and by August 9, 23,000 men had been taken prisoner. The balloons of Bz 6, 56, 74, 50, 137 and 221 naturally ascended immediately at the start of the attack, but the telephone lines to the balloon control center were largely destroyed by enemy artillery. As a result, only a few reports reached the balloon control center in Foreste. The balloon launch sites had been under artillery fire and machine-gun fire from airplanes since the morning. After a few hours, the Allied advance forced the balloon trains to march back. The balloon of Bz 72 was attacked twice by enemy aircraft while in transport (balloon up). During the first attack, the

Above: American memorial in Flirey commemorating the Battle of St. Mihiel on September 12, 1918.

Above: The German balloons were only rarely able to ascend undisturbed in August 1918.

observer was shot in the stomach and during the second attack, the balloon was badly damaged. Finally, the observer bailed out after he had deflated the balloon envelope with the tear track. Immediately afterwards, the first enemy tanks appeared. The balloons of the Bz 6 at Caix and Bz 50 at Proyart were burnt after air attacks. The MG of Bz 50 shot down a Sopwith Camel at around 9.45 am.

Attacks on Bz 56, 74, and 221 were prevented by the anti-aircraft guns. The balloons of Bz 25 and 137 were destroyed by incendiary bombs and the headquarters of FLA 5 was bombarded by enemy aircraft and machine-gun fire.

The balloons of the FLA 29 and 37 of the 18th Army took off and were also immediately attacked. The balloon of Bz 118 was hit by several artillery shells and Bz 127 was set on fire by two aircraft. Only the anti-aircraft gun of Bz 83 managed to fend off the enemy aircraft.

Although the Allied aircraft did not yet fully engage in combat on August 8, German losses were considerable. On August 9, the weather improved and nine German balloons observed the area of the 2nd Army from 6:00 am to 10:00 pm. Heavy enemy artillery was reported at Cérisy, Vauvillers, Rouvroy and Harbonnières. Five balloons flew over the 18th Army from 5.00 am to 9.45 pm. Eighteen enemy batteries were detected and six of them were fired upon. The attacks on the balloons decreased noticeably after German fighter planes appeared more frequently at the front. At 7.20 a.m. six Spad attacked Bz 74 without success. At 18.25 the second attack on Bz 74 took place, during which a French Spad XIII was shot down by German aircraft. At 8.45 a.m. it was the turn of Bz 151, but this attack was also repulsed. However, some flak fragments hit the balloon envelope. At midday, the balloon of Bz 151 exploded near Roye after the attack by Lt Marcel Haegelen of Spa 100 and 1Lt Stephenson.

The next day, the 2nd Army observed 7 balloons from 6.00 am to 9.50 pm. In the 18th Army 5 balloons went up, the Bz 83 and 119 were attacked several times by airplanes and artillery. In the following days, too, all available balloons of both armies went up. Bz 119 near Champien was the target of three Spad from 19:15 to 19:30. The balloon remained undamaged, but the machine gun fire killed several soldiers on the ascent site.

Although the front was stabilized again by 18 August, the German army in particular had lost faith in victory or a peace agreement. The army had suffered such heavy losses that Hindenburg and Ludendorff also called on the Kaiser to negotiate peace.

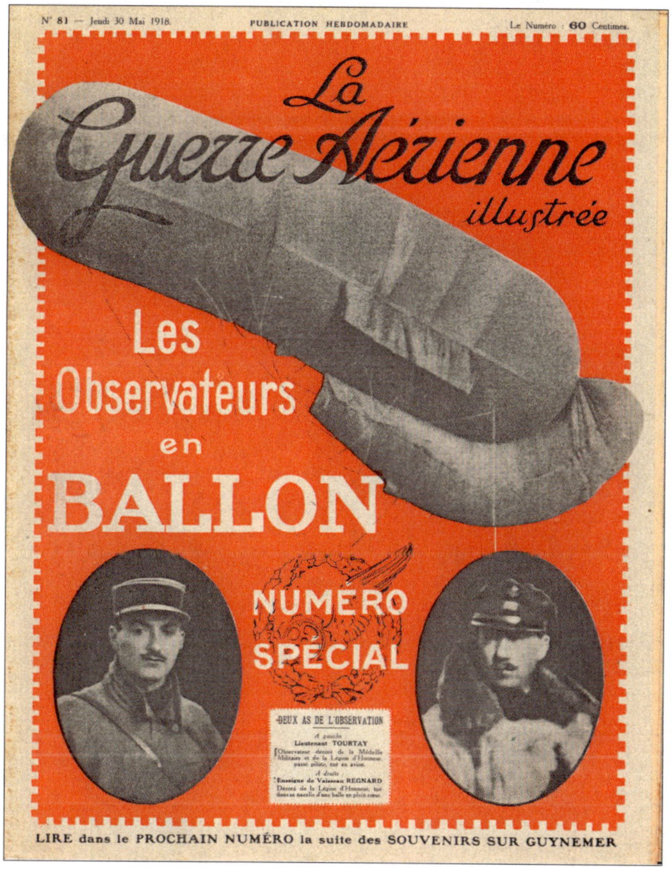

Above: In May 1918, the newspaper *La Guerre Aérienne* dedicated a special edition to the balloon troops.

Above: A French balloon observer with a parachute harness. The parachute is attached to the outside of the basket in a container.

Hanriot HD.1 No.23
Lt. Willy Coppens
37 total victories, 35 balloons
Coppens was the leading Belgian ace and the leading balloon buster of WWI.

The American Meuse – Argonne Offensive

Above: In September 1918, the 1st US Army moved unnoticed into the Argonne. The Balloon Companies also spread out in the wooded area without being detected by German reconnaissance.

Above: On September 26, 1918, two US balloons take off near Montfaucon.

With American President Wilson's declaration of war on the German Reich on April 6, 1917, a new opponent and thus also the world's most important economic power took up arms against Germany. In addition to the urgently needed raw materials, the Entente also received an influx of well-trained and combat-ready soldiers. Under General John Pershing, the Americans demonstrated their military prowess on September 12, 1918 in the Battle of St. Mihiel and from September 26 in the Meuse-Argonne Offensive.

By mid-September 1918, the 1st US Army, which consisted of 14 divisions, had completed its deployment on the Argonne front. General Pershing also brought in strong tank divisions and aviation forces to support the infantry. These included the 1st Pursuit Group, which consisted of the 94th, 95th, 27th and 147th Pursuit Squadrons and the 218th Aero Park Squadron, later the 4th Aero Park Squadron.

From the German side, the Argonne, which was part of the 5th German Army, was considered a quiet front area. After the German attack on Verdun, a few kilometers to the east, had been halted in 1916.

From September 3, 1917, the 2nd Württemberg Landwehr Division held the entire Argonne front. It was not until the beginning of September 1918 that the 27th Inf. Div. (Württ.) with the Geb. Regiment and the Feld Art. Rgt. 49, the 1st Guards Infantry Division and the K.u.K 1st Infantry Division in the Dun sur Meuse area. In the fall of 1918, the German air force of the 5th Army consisted of Fliegerabteilungen 36, 44, 274, 280 and Jagdstaffeln 65, 67 and 68. The air force was only reinforced after the start of the American offensive. The following airship units were assigned to the 5th Army: Stab bayr. Feldluftschiffer-Abt. 66, Stab Feldluftschiffer-Abt. 12 u. 31 as well as the balloon platoons:

Above: Uffz. Hans Marwede immediately after his emergency landing. (US Signal Corps)

11, 12, 34, 35, 64, 81, 95, 109, 115, 121, 129 and 216.

On the night of September 25/26, 1918, the American offensive in the Argonne began with the usual barrage lasting several hours. After daylight, the US infantry, supported by tanks and aircraft, went on the attack. Over the next 22 days, the German troops were gradually pushed back in fierce fighting until the front had stabilized again on 17 October 1918. Fierce air battles took place during the battle, with the balloon troops also suffering heavy losses on both sides. In

Above: The OAW Fokker D..VII 4092/18 of Uffz. Marwede after his emergency landing on October 3, 1918 (US Signal Corps)

Above: US soldiers examine the wreckage of Marwede's Fokker D.VII (Greg VanWyngarden)

Above: Hans Marwede after his capture by the US soldiers. (Greg VanWyngarden)

Above: Uffz. Marwede appears to be in shock; he sits thoughtfully next to the US soldier. (Greg VanWyngarden)

Above: The fabric of the Fokker D.VII was cut open to free Marwede from the wreckage. (Greg VanWyngarden)

Above: Marwede's Fokker D.VII (OAW) 4082/18 (Greg VanWyngarden)

the partly wooded and often foggy terrain of the Argonne, enemy observation and the firing of artillery by tethered balloons was particularly important, which is why the balloons were among the most important targets for artillery and fighter pilots on both sides.

Private Fritz Ibrügger was captured by the Americans in the Argonne on September 26. Ibrügger's listening station 215 was located on the Gercourt - Béthincourt road in an underground bunker, from which the French telephones were tapped using underground cables. Since September 24, however, the French remained silent and no more telephone conversations could be intercepted. As an experienced soldier, Ibrügger knew that something was wrong. Meanwhile, on the other side of the front, 600,000 American soldiers from the 1st Army had been exchanged for French soldiers without anyone noticing. On the morning of September 26, the American troops overran the German lines and also captured the crew of listening station 215. In German, the Americans sent Ibrügger and his comrades towards the American lines. They were collected at Esnes; attempts by American soldiers to obtain a souvenir were immediately stopped by the US officers. Ibrügger continued to write:

"We are approaching the line of enemy tethered balloons. In a small section there are 10 of these sausages hanging in the sky. [...] Then a call that jumps through the (POW) column like an electric spark. German airman! The Americans take notice. Their defense batteries and machine guns start a wild shootout. They know the airman's target, it's the balloons.

"Only a few hundred meters separate him from his target. Now the observer jumps out with his parachute. A short rattle of the machine gun and the balloon goes up in flames. A tremor runs through the plane and it slowly descends. There is a roar of cheers from the American columns on the street. But suddenly the whirring of the propeller becomes stronger again and it rises again. Slowly our plane disappears towards the German front."

The American balloons of the 8th Balloon Company had risen in the Montfaucon - Romagne-sous-Montfaucon area. The observers, 2Lt HB Hudnot and 2Lt CJ Ross, bailed out after the ground troops gave the order. While Hudnot immediately complied and landed safely, Ross hesitated too long. After the jump, his parachute was caught by the burning balloon and he fell to his death from a height of 300 meters. Cleo Ross was the only observer of the Balloon Service to be killed in action. The German Fokker D.VII was probably flown by Lt. Hebler of Jasta 68.

On October 3, 1918, NCO Hans Marwede of Jasta 67 took off with the OAW Fokker D.7 4092/18 on a fighter flight on enemy balloons. A few days earlier, on September 14, Marwede had shot down three French balloons of the 25th, 30th and 31st Compagnie d'Aérostiers near Fromerville within a few minutes. The observers, SLt Boret, Adj André Lurcat and Cpl Guilbert parachuted out. In the afternoon of October 3, Uffz. Marwede sighted the balloon of the 6th US Balloon Company near Montfaucon. The observers, 1Lt J R Taylor and a sergeant, immediately jumped out with the parachute. Seconds later, Marwede set the balloon on fire. Now the machine guns of the 6th US Balloon Company and the infantry opened heavy fire on the Fokker, and Marwede, hit in the engine, had to make an emergency landing in a nearby field. The aircraft overturned and came to rest on its back. The Americans, who had run up, freed the trapped Marwede and took him prisoner. During the emergency landing, Uffz. Marwede suffered only minor injuries to his face and after an American cigarette he resigned himself to his fate as a POW. By chance, film reporters from the US Signal Corps were at the scene

Above: The crash site of Uffz. Marwede on today's D 18; in the background on Hill 336 the American Memorial of the US First Army, which commemorates the Meuse-Argonne Offensive.

Above: The church at Montfaucon was not rebuilt after the war. In the foreground, a former bunker of the German troops.

documenting the event. The US soldiers cut large pieces of canvas out of the wreckage of the Fokker D.7 as a souvenir before it was taken away.

At the end of September 1918, the French and British armies also launched their final offensives against the German army in Flanders, near Arras and at Cambrai. The severely weakened German army was no longer able to counter this. On September 28, Hindenburg and Ludendorff decided to enter into direct peace negotiations based on President Wilson's 14 points. After the armistice on 10 November 1918, the field airmen also marched home to the six replacement divisions and were demobilized. With the dissolution of the office of the Commanding General of the Air Force on January 16, 1919, the history of the German field airship troops also came to an end. The German field airships lost 471 balloons to Allied aircraft. In addition, a further 75 balloons fell victim to enemy artillery and 109 balloons were burned in accidents. By the end of the war, 614 enemy balloons were recognized as air victories for the German airmen.

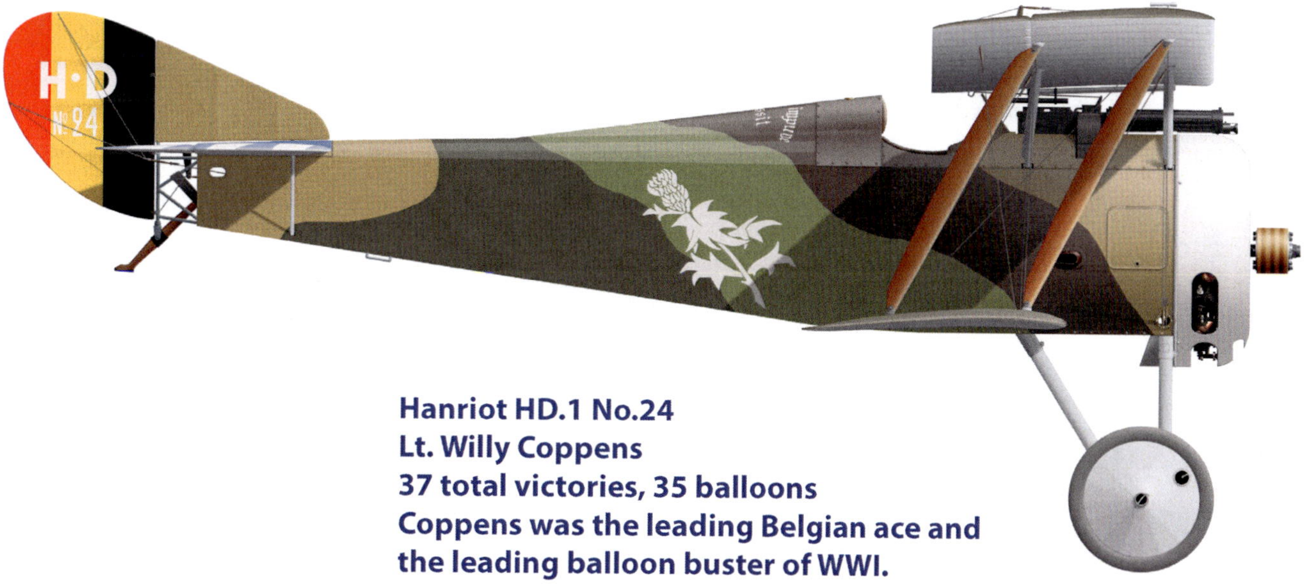

Hanriot HD.1 No.24
Lt. Willy Coppens
37 total victories, 35 balloons
Coppens was the leading Belgian ace and the leading balloon buster of WWI.

Attacks on Balloons

Above: The Nieuport 17 of the N68, flown by Maréchal de Logis Blaise Préher, was forced to land by Lt. Gebhardt Saltzwedel of the Jasta 24 on April 11, 1917. On the lower wings, the sheet metal fairings on the V-struts can be seen, which prevented the risk of fire from the Le Prieur rockets.

Above: MdL Marcel Careé of the N 112 attacked a German balloon near Leffincourt on April 12, 1917, where he came under fire from the ground machine guns and had to make an emergency landing with the engine stopped.

As the war progressed, all parties recognized the importance of balloons as the "eyes" of the infantry and artillery. While only three balloons were destroyed by enemy action on the German side up to 1915, the losses rose to 37 in 1916. The German air force, the French Aviation Militaire and the RFC fought the enemy balloons with the usual steel or lead-core projectiles up to 1915. The destruction of a balloon by aerial arrows or incendiary bombs was rather accidental. The German standard lead-core ammunition (K ammunition) in 7.92 mm caliber proved to be mostly ineffective; the K ammunition penetrated the balloon envelope and left only small holes with minimal gas loss. The first German balloon victory was achieved on October 14, 1915 by the leader of FFA 70, Hptm. Ernst Krüger. The French balloon ignited after Krüger fired at it with a signal pistol. The explosive ammunition used by the Austrian Alder company in 1914 still had to be produced at great expense. In the balloon cartridge (B cartridge), a powder charge exploded on impact and was intended to ignite the gas in the balloon. However, it was not until the easy-to-produce phosphorus cartridges were introduced that the balloons could be effectively combated. The 7.92 mm projectiles were filled with yellow phosphorus, which ignited in the air. As the projectile heated up in the MG barrel, a hole filled with light solder opened in the projectile and the phosphorus escaped. The escaping, burning phosphorus then set fire to the gas in the balloon or the gas-air mixture on the damaged balloon envelope. The manufacturers of this special ammunition were the companies: Fabrik für Feinmechanik, Carl Lindström Berlin, Farbwerke Höchst and Chemische Fabrik Merck in Darmstadt. As the phosphorus ammunition caused severe poisoning, even with grazing shots, the Kogenluft only permitted its use to combat enemy balloons and had to be approved in writing by the squadron leader.

Each pilot had his own "recipe" for the "mixture" of K and phosphorus ammunition in the MG belts. Lt. Paul Bäumer of the Jasta 5 had one MG belted with phosphorus ammunition and the other with K ammunition. On July 13, 1917, Lt. Bäumer shot down a balloon of the 45th Compagnie d'Aérostiers near St. Quentin:

"At close range, I fired 10 to 15 rounds of phosphorus ammunition into the large target. Then the left machine gun failed, the firing pin broke. I kept firing K ammunition with the right machine gun until I was within 10 meters."

On August 12, 1917, the Jasta 26 was ordered to attack enemy balloons at Vlamertinghe at 5:30 pm. From 5.20 to 5.40 p.m., its own artillery took the balloons' protective anti-aircraft guns under fire, so that the Jasta 26's aircraft had little to fear from the anti-aircraft guns. After taking off from Iseghem airfield, Lt. Xaver Dannhuber, Lt. Werner Dahm and Vzfw. Otto Fruhner flew over the front at 17:20. While Dahm and Fruhner took over the cover, Dannhuber flew at top speed towards the British balloon:

"I haven't fired a shot yet when I'm presented with a spectacle I've never seen before. Two ever-increasing white dots detached themselves from the balloon. The two observers had jumped off, realizing the imminent danger. I fire a series of 50 shots at the giant. I thought I was about to ram it and pulled up my machine. Assuming that the thing would burn up immediately. Deception! The balloon remains unperturbed. I attack again. Now small flames are flickering out of about 20 wounds and getting bigger and bigger. It's on fire."

On August 1, 1918, five aircraft of the Jasta 46 took off from Moislains airfield to attack enemy tethered balloons. The deputy officer Robert Heibert and Lt. Helmut Steinbrecher of the Jasta 46 had their MGs belted with phosphorus and K cartridges (3 phosphorus - 2 K shells):

Right: Lt. Böhler, 2nd from right in front of his Fokker D.VII (OAW).

"On August 1, 1918, at 5:40 in the afternoon, my chain of five planes took off on a fighter flight. I climbed to 5000 meters above the airfield and then flew to the front. Since I had convinced myself by flying over the front twice that there were no enemy fighters, I decided to attack the enemy balloon at Bussy-les-Daours. The other tracked fighters immediately understood what I intended to do.

"I flew over the balloon at an altitude of 5,000 meters without being hit by enemy anti-aircraft fire. I then switched off the engine and descended in steep spirals and dives towards the balloon at Bussy-lés-Daours. The chain followed closely up to an altitude of 2500 meters. Vice Sergeant Hennrich went down with me to 1000 meters on the balloon.

"I shot at the balloon at close range and a short time later it was on fire. The next balloon near Aubigny, west of Bonnay, was attacked by Vzfw. Hennrich and I attacked it together and set it on fire.

"As I was convinced that there were still no enemy fighters to be seen, I decided to attack the balloon near Daours again despite the strongest barrage. Vzfw. Hennrich also attacked a balloon further south near Blangy-Tronville. Both balloons burned at about the same time. We attacked the fifth balloon near Gentelles both together. The balloon did not burn despite being approached several times. The two observers jumped out in parachutes, as with the previous ones. When I looked around again, I could see all the observers jumping off all along the front as far as Montdidier. I saw the white dots of the parachutes shining brightly in the sun. All the balloons on the English front were withdrawn.

"The three remaining chain members, who had the task of covering the attackers' backs, were involved in a dogfight with four enemy two-seaters. One of them was forced to land.

"We had to fly back over the front at the lowest altitude (200 meters) in the strongest enemy fire from flak and machine guns."

The British pilots of No. 40 Squadron RFC had practiced flying low to attack balloons with their Nieuport 17s in order to surprise the German balloon defences. On May 2, 1917, they attacked the four balloons of the Bavarian FLA 1 (Bz 1, 2, 3 and 4):

Above: Hermann Böhler of Jasta 63.

"Lieut. K. Mackenzie and 2nd-Lieuts. Walder, Lemon, Thompson, Bond and Morgan, No 40 Squadron, crossed

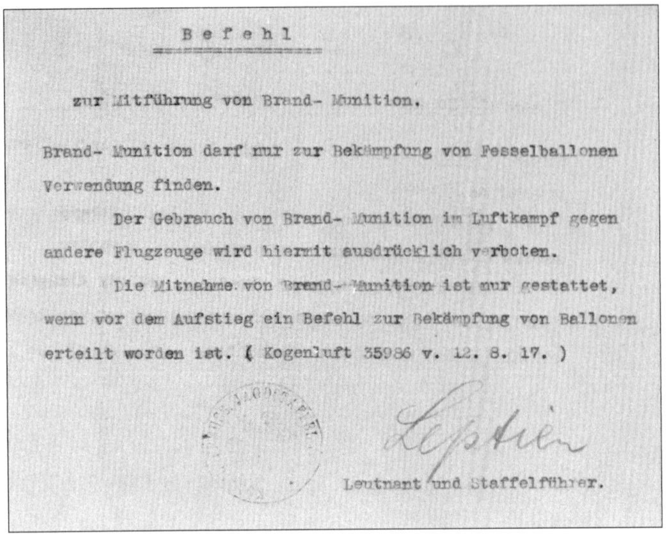

Befehl

zur Mitführung von Brand- Munition.

Brand- Munition darf nur zur Bekämpfung von Fesselballonen Verwendung finden.

Der Gebrauch von Brand- Munition im Luftkampf gegen andere Flugzeuge wird hiermit ausdrücklich verboten.

Die Mitnahme von Brand- Munition ist nur gestattet, wenn vor dem Aufstieg ein Befehl zur Bekämpfung von Ballonen erteilt worden ist. (Kogenluft 35986 v. 12. 8. 17.)

Leptien

Leutnant und Staffelführer.

Above: Instruction issued by Kogenluft on August 12, 1917, regulating the use of incendiary ammunition.

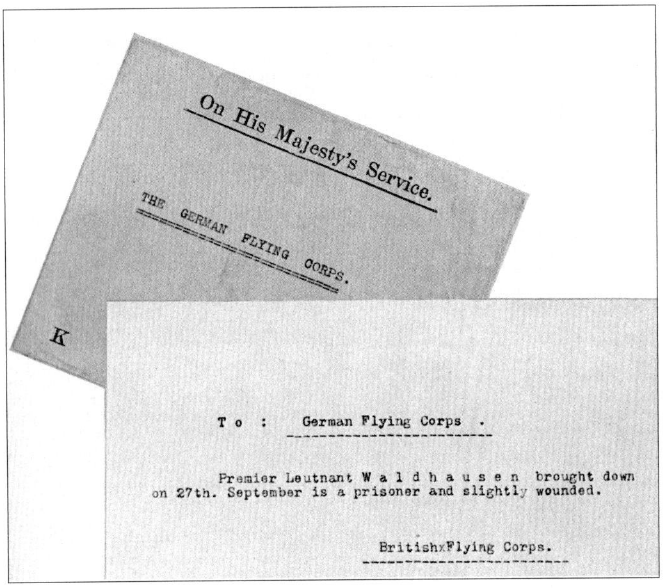

On His Majesty's Service.

THE GERMAN FLYING CORPS.

K

To : German Flying Corps .

Premier Leutnant W a l d h a u s e n brought down on 27th. September is a prisoner and slightly wounded.

BritishxFlying Corps.

Above: Lt. Hans Waldhausen of the Jasta 37 was defeated by British balloon protection after his third balloon victory on September 27, 1917. Lt. JH Tudhope of No. 40 Sqn and Flight Commander CD Booker of No. 8 Naval Squadron.

the lines at 50 feet and under in order to attack German balloons. Previous experience had shown the necessity of flying low in order to obtain satisfactory results in these attacks.

These pilots had practised low flying, and had learned how to use trees, houses and the ground in screening them from infantery and machine gun fire. Artillery co-operated by putting a heavy barrage on the German trenches, leaving the pilots a certain aera in which to crows. As a result of the attack four German balloons were destroyed."

The Inspection of Aviation, Munich, received the following report from the Bavarian FLA 1:

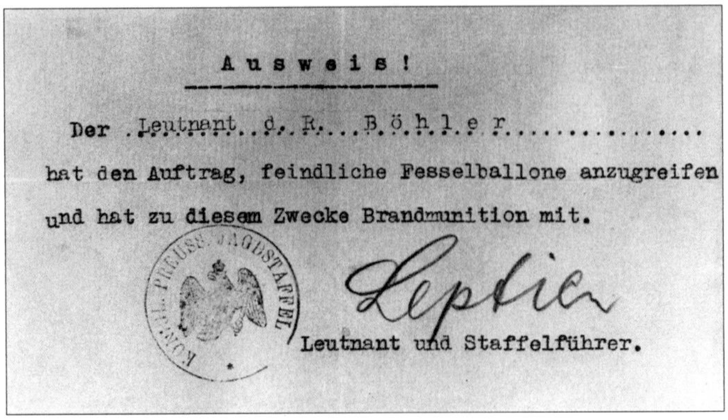

Ausweis!

Der Leutnant d. R. B ö h l e r ....

hat den Auftrag, feindliche Fesselballone anzugreifen und hat zu diesem Zwecke Brandmunition mit.

Leptien

Leutnant und Staffelführer.

Above: ID from Lt. Böhler, who authorized the phosphorus ammunition.

"On May 2, 1917, at 10:30 a.m., four enemy fighter planes carried out a single attack on all four balloons of the division. The four planes flew along the front at a conspicuously low altitude of between 300 and 400 meters and suddenly turned onto the balloons. The balloons, which were at an altitude of about 680 meters, could be caught up to 300 meters and in some cases completely.

One balloon, bayer. Bz 2, was attacked by the first airman at a distance of 300 - 400 meters with incendiary ammunition and later from very close range. The aircraft flew over the balloon at a very low altitude. The second plane turned in a south-westerly direction shortly before the balloon without bothering the balloon. This plane seems to have had the task of protecting the first plane.

The third and fourth aircraft attacked balloon train 4. While the incendiary ammunition at bay. Bz 2 destroyed the balloon B. A 5 R it was not used on bay. Balloons 3 and 4 did not ignite.

The balloon observers of bayer. Bz 2, Lt. d.R. Schneider and from bayer. Bz 4, Lt. d.R. Walch, jumped with a parachute and landed. After the jump, Lt. d.R. Walch was shot at again with machine guns by the airman."

All observers suffered only minor wounds. The Nieuports of No 40 Squadron were shot at with machine guns and anti-aircraft guns, but escaped due to their low altitude. The balloon of Bz 2 was completely burnt up. The balloons of Bz 3 and Bz 4 suffered a few machine gun hits, but could be recovered. The balloon of Bz 1 was not attacked.

The French fighter planes had already been securing their tactically important balloons since 1915. Lt. Erik Thomas, experienced with 9 balloon launches, also had to make this experience. As a member of the Jasta 9, Vzfw. Thomas achieved his first aerial victory over a balloon of the 67th Compagnie d'Aérostiers near St. Jean in the 3rd Army in Champagne on January 3, 1918. This was followed on January 12 by the next and on February 2 by a balloon of the 57th Compagnie d'Aérostiers. Meanwhile promoted to lieutenant, Thomas shot down a balloon of the 87th Compagnie d'Aérostiers near Somme-Suippes on March 6. The two balloon observers, SLt Gustave Lemaitre and

Above: The deputy officer Robert Heibert of Jasta 46 was awarded the Golden Military Cross of Merit on July 5, 1918. This highest decoration was also awarded to Vzfw. Oskar Hennrich on November 3, 1918. (B. Schmäling)

Helmut Steinbrecher of Jasta 46 (B. Schmäling)

Above: The officer's deputy Robert Heibert and Lt. Helmut Steinbrecher on the airfield of Jasta 46 (B. Schmäling)

1Lt Sidney Howell of the US Balloon Service, parachuted out. After his training at the AEF Balloon School in Cuperly, Howell had been assigned to the 87th Compagnie d'Aérostiers for front-line training. He returned to Cuperly at the end of March and served in the 1st Balloon Company of the AEF from April 1918 until the end of the war in the St. Mihiel and Meuse-Argonne battles. Lt. Thomas increased the number of his balloon victories to 9 by March 23.

Transferred to Jasta 22 on March 17, 1918, Thomas was assigned to fight the French balloons in the area of the 18th Army as part of the German March Offensive. On March 23, Squadron 22 had just returned from the front when Thomas noticed a distant balloon:

"I immediately turn back to the front and, as there is no enemy balloon in front of our army, I attack the next balloon in front of the neighboring army. I push off at full throttle and have just started firing at it with phosphorus ammunition when there is a rattling behind me and a French fighter plane fires at me. I quickly turn back and the two of us have a prolonged cornering battle. But the Frenchman has the better plane. The Frenchman gets closer and closer to me and I have to see how I can get out of the enemy's field of

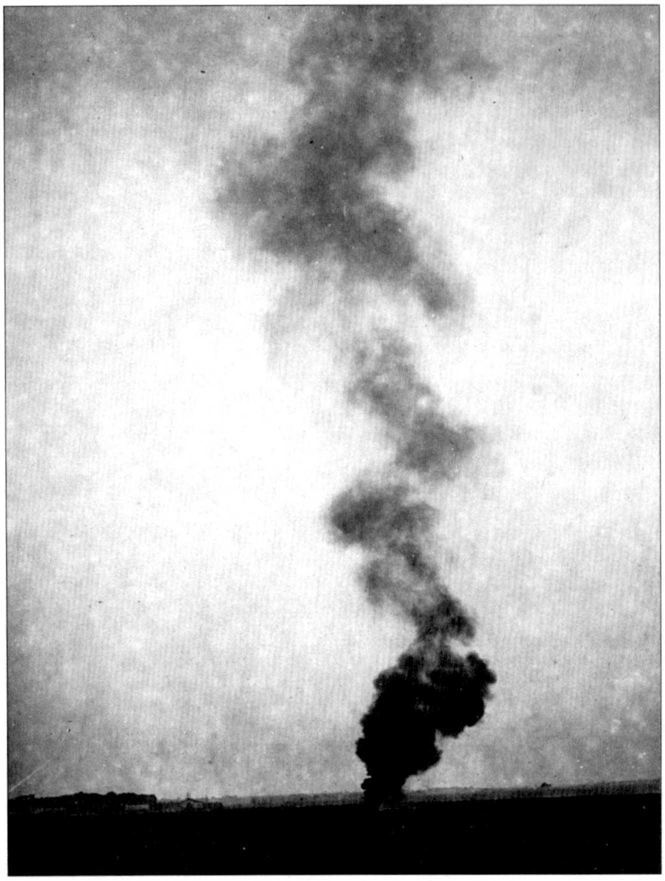

Above: The burnt envelope of a French balloon falls to the ground. (Greg VanWyngarden)

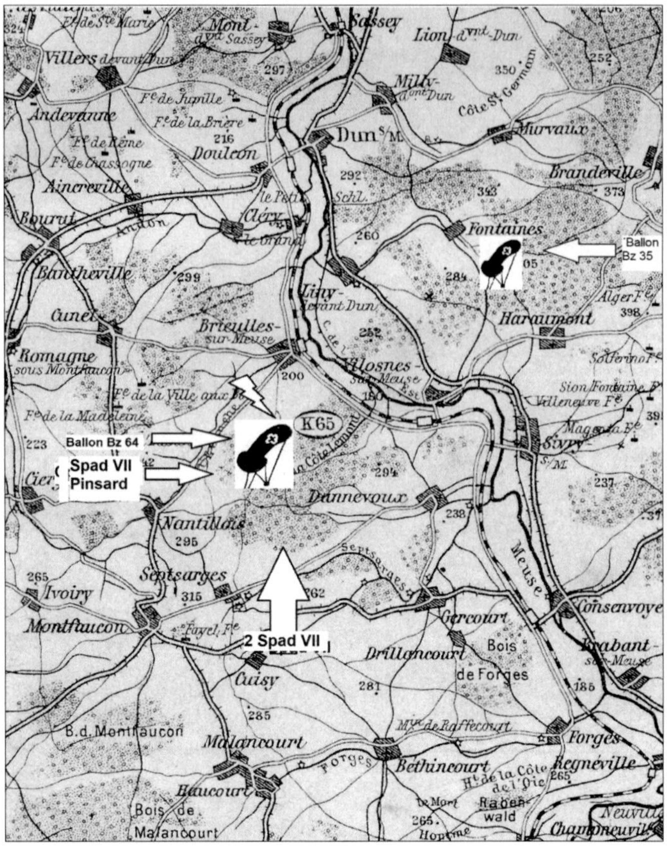

Above: On May 4, 1918, the Spad VII of Spa 23 attacked balloon platoons 35 and 64. Captain Pinsard shot down balloon No. Ae XXII 1203 B.H. of Bz 64.

fire. I'm under his wings in a flash so that he can't see me. I'm just about to start shooting at him when there's another rattle and crash behind me. Another Frenchman was behind me. My controls stop responding and the plane crashes vertically from a height of 300 to 400 meters."

Lt. Thomas was taken prisoner by the French, from which he returned home in 1920. The Lt. Unger accompanying him was also shot down, a third pilot of the Jasta 22 was able to escape. The opponents of Lt. Thomas and Lt. Unger were Lieutenants Jean Chaput and August Laholle of Escadrille Spa 57.

The balloons were retracted as quickly as possible by the crews during enemy aircraft attacks and so the action often shifted to ground level. Here, the attacker came into the effective range of the MG and therefore tried to reach their own lines again immediately after the attack by flying low. With brightly painted aircraft and clear skies, attacking balloons was a hopeless undertaking. Vzfw. Hans Brzenk, a member of the Jasta 67, which was based near Marville in Lorraine in March 1918, described the preparations for a balloon attack:

"For several days I had been observing through the scissor telescope the long line of French tethered balloons about 12 kilometers behind the French front at altitudes of 500 to 1500 meters, surrounded and guarded by enemy fighter planes. Twice I had already attacked such a gas bubble (balloon), but each time they pulled it in with monkey-like speed and laid a murderous barrage in front of my nose with their anti-aircraft batteries. [...] But one day the weather was favorable. It was very hazy, the sky slightly overcast. The clouds stretched out in little balls and heaps at an altitude of 1200 meters. I felt lion's courage in my chest. I put on fresh clothes and my best uniform. After all, the danger of being captured by the enemy is particularly great in a captive balloon and then it is especially unpleasant not to appear as a gentleman.

"After flying around the front for a while, I dived over the front towards the highest balloon at about 1200 meters. I had already covered three quarters of the distance when suddenly countless black and white clouds (flak shots) appeared behind me. [...] In a steep dive I turned my machine downwards. Now I am within 100 meters. The tracer and phosphorus ammunition disappears in long white threads of smoke in the balloon decorated with the blue and white cockade. I race close over the gas bubble and look around. Both observers parachute down one after the other, but the black monster doesn't burn."

Brzenk was able to approach and attack his target by taking advantage of the clouds, but Phosphorus ammunition failed to ignite the balloon even after three approaches. At

Above: In August 1917, a British Bristol Fighter tried in vain to shoot down the balloon of FLA 34. After the observer, Lt. Franz Schneider, had parachuted out, the anti-aircraft gun opened fire and was able to shoot down the British fighter.

the same time and with numerous hits in his Pfalz D.IIIa, Brzenk escaped at a height of 20 to 30 meters above his own lines. Even for experienced fighter pilots like Oblt. Hans Waldhausen of the Jasta 37, the balloons represented a great risk. Waldhausen had already already shot down two balloons when he took off on September 27, 1917 to attack a French Saucisse on September 27, 1917. After he had shot down the balloon, he was attacked by British balloon guards, Lt. JH Tudhope of No. 40 Sqn and Flight Commander CD Booker of the No. 8 Naval Squadron, attacked and forced Waldhausen to make an emergency landing near Souchez.

On the evening of May 4, 1918, the balloon (No. A XXII 1203 B.H.) of Bz 64 stood south of Brieulles-sur-Meuse at an altitude of 700 - 800 meters above the ascent site. The observers Lt. Kuhna and Luftschiffer Hübner were practicing flashing signals when Maschinenflakzug 77 reported two French aircraft. The Spad VIIs of SPA 23

Above: Lt. Schneider evaluating circular images.

belonged to Groupe de Combat 22 and had taken off from Souilly towards the front under the command of Captain Armand Pinsard. The formation consisted of four Spad VIIs, two of which were intended to deflect German anti-aircraft fire. Machine-gun platoon 77 also immediately warned the balloon crews with three warning shots. Meanwhile, the two other Spad VIIs approached the German balloons of Bz 64 and Bz 35 undetected from an altitude of 4000 meters in gliding flight. 700 meters before the balloon of Bz 64, the Spad VII of Capitaine Pinsard was spotted. The five machine guns and two automatic cannons of the Bz 64 immediately fired tracer ammunition at the Spad. Pinsard began firing at the balloon at a distance of 300 meters. After about 50 shots, Pinsard pulled his Spad a few meters above the balloon, flying through the barrage of machine gun and anti-aircraft fire. Shortly afterwards, the balloon exploded. The observers, Lieutenant Kuhna and Airman Hübner, were able to bail out in time.

According to the balloon crews' observations, Captain Pinsard's Spad VII must have been hit by anti-aircraft fire (120 rounds of 3.7 cm flak and 600 rounds of machine gun fire). However, all the aircraft were able to escape over the front again. The simultaneous attack on the balloon of Bz 35 near Fontaines was repulsed by the Maschinen-Flakzug 53.

Very rarely did the balloon observer manage to fend off an enemy from the basket with a pistol or carbine. While working in the balloon, the observers naturally concentrated entirely on their mission and there was little time to observe the airspace. In addition, aiming a rifle from the swaying basket was almost impossible. On the morning of November 9, 1916, Lt. d.R. Hepner of the bay. FLA 8 at Meurchin as usual and observed the enemy hinterland. At 10.30 a.m. he noticed a British aircraft a few kilometers away. Hepner could clearly see the cockades on the wings through his binoculars. He immediately informed the ground crew

Above: A 3.7 cm Maschinenflak mounted on a Krupp truck, with the rangefinder (left) targets up to 5000 meters could be measured.

and loaded his carbine. The British FE8 (7624.6) of No 40 Squadron steered Capt T Mapplebeck along the front for some time and then turned to attack Hepner's balloon. Now Lt. Hepner fired his carbine several times at the FE8, hitting the front fuselage of the aircraft and wounding Capt Mapplebeck. Shortly afterwards the balloon caught fire and Hepner jumped out of the basket at 10.45 am. Mapplebeck had to make an emergency landing of the FE8 in a field near Provin and was taken prisoner. On December 11, 1916, Lt. Hepner received a letter of commendation from Kogenluft. Thanks to his courageous conduct, the FE8 was forced to land and Capt. Mapplebeck was captured.

Aircraft that were used specifically for balloon hunting were often given a special camouflage livery. Low-hanging clouds were often used as cover for the aircraft to approach the balloons, which were heavily protected by machine guns and anti-aircraft guns, as unnoticed as possible. This was because firing at the balloons with phosphorus ammunition was only successful from a distance of 20 to 30 meters. Some Albatros D aircraft in the Jasta 15 were also given a special camouflage livery. Lieutenant Heinrich Gontermann had the wings and fuselage of his Albatros camouflaged in gray-blue with gray-green patches so that he could fly as close as possible to the balloons under cover of clouds. Gontermann's Albatros D.V also had a red fuselage band and two red, white and black bands on the tailplane. This served on the one hand as recognition for his own trench crews when he crossed the front line again, and on the other hand to confirm his aerial victories by independent witnesses.

Not only fighter planes chased enemy balloons. Lt. Guthoff of Flieger-Abteilung 42 shot down a total of four balloons with a C-plane in the Southern Army (9th Army). On July 30, 1917, Guthoff scored a double near Ivesti during a fighter flight. A few days later, on August 8, Lt. Guthoff did not return from the enemy flight.

The RFC's No. 17 Balloon Section made an exception in order to fend off or kill the enemy fighter pilot during the balloon attack. On November 21, 1917, Lt. Rudolph von Eschwege attacked a British balloon on the Struma front near Orljak. As a member of FA 30, Eschwege had shot down balloons of the No. 17 Balloon Section on October 28, November 9 and 15, 1917. On November 21, a British balloon was exceptionally high, at least 800 meters according

Above: Vzfw. Hans Brzenk, member of Jasta 67.

to Bulgarian reports. Eschwege approached the balloon with his Albatros D.III and fired into the envelope. As he flew over the balloon, it exploded with tremendous force. The British had detonated a hidden 350 kg ecrasite charge in the basket, which severely damaged Eschwege's Albatros and caused him to crash fatally.

In the course of the war, some fighter pilots specialized in enemy balloons.

The list of balloon fighters is headed by the Belgian Lieutenant Willy Omer Francois Coppens, Chevalier de Houthulst. Of his 37 aerial victories, 35 were against German balloons. He often shot down two or three balloons in one day. On October 14, 1918, Coppens was seriously wounded by anti-aircraft fire when he achieved his 37th aerial victory over a German balloon near Thourout.

He was followed by the Frenchmen Léon Boujarde (27 balloons), Michel Coiffard (24 balloons) and Maurice Boyau (21 balloons). Oberleutnant d.R. Friedrich Röth shot down 20 enemy balloons, making him the most successful German balloon fighter. He achieved three balloon victories on January 25 and four on April 1, 1918. The 5th balloon was shot down by Vzfw. Gilbert Wagner shot down the fifth balloon. On May 29, 1918, Röth even managed to shoot down five balloons in a row at Jasta 16b:

Above: Lieutenant Chaput with the wounded Lt. Unger, front, and Lt. d.R. Thomas in French captivity. (*Le Miroir*, May 19, 1918)

Left: Sous-Lieutenant Jean Chaput of Escadrille Spa 57 was promoted to lieutenant on March 25, 1918.

"On May 29th, I saw ten English tethered balloons from our airfield at 3 o'clock in the afternoon. After I had been thinking about an attack on the line of balloons on this front for three weeks, the conditions seemed favorable to me for the first time that afternoon. There were no English fighter squadrons to be seen. Only a few working airplanes approached the front from time to time. I took off alone at 3.30 p.m. and flew against the wind over Worvik and Roeselare.

Left: Also in the basket of the 87th Company balloon on March 6 was 1Lt. Sidney Howell of the US Balloon Service. Both observers parachuted out.

Above: Lt. d.R. Hepner in front of the British FE8 (7624) of No. 40 Squadron at Provin.

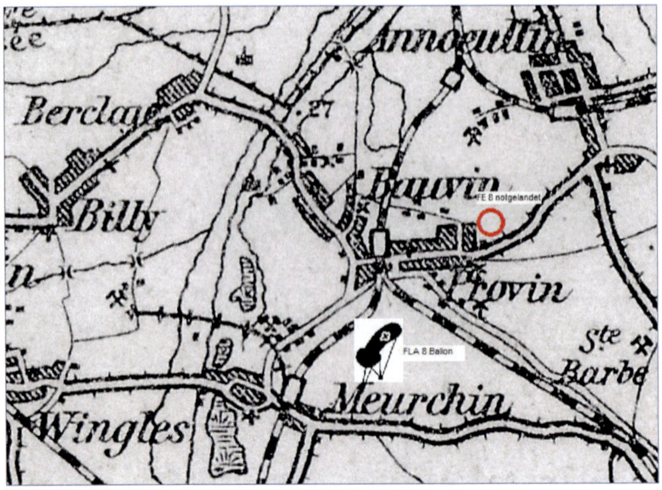

Above: Map of the location of FLA 8 near Meurchin and Provin, the landing site of Capt. Mapplebeck.

"South of Diksmuide I drifted into English territory at an altitude of 2000 meters. In the furious fire of the English earth machine guns, I shot down three tethered balloons between Diksmuide and Poperinge and two between Poperinge and Hazebrouck.

"Two observers jumped out of each of the two balloons that were shot down, and seven observers jumped out of the five balloons that crashed. One observer jumped from a tethered balloon that was too low for my attack at an altitude of about 300 meters. From another balloon, which was about 500 meters higher than I was flying, two observers jumped from above. All in all, ten observers in the English balloon line had jumped out of the balloon baskets with their parachutes.

"Shortly after the third launch, a twin-winged aircraft with French cockades, probably a Spad, attacked me southwest of Poperinge. I escaped its attack by turning in the opposite direction. When I managed to get behind him, he pushed into a right turn. This gave me a lead that he could no longer catch up with. When I had shot down five captive balloons, three Englishmen approached me from Bailleul. I therefore gave up my attack on two balloons that were still high northeast of Hazebrouck. I turned back to our

Above: The probability of a successful defense against an enemy aircraft with a carbine was very low.

Above: Lt. Hepner's balloon burns up on the ground.

Above: A thoughtful Lt. Hepner in front of the burnt remains of his balloon.

Above: One projectile hit the fuselage of the FE8, the number G37 on the cockade is unknown.

Above: The FE8 near the village of Provin.

Right: The letter from Kogenluft to Lt. d.R. Hepner dated December 11, 1916.

Abschrift.

Der kommandierende General
der Luftstreitkräfte
Nr. 35078 L.

47

Gr. H. Qu., den 11. 12. 1916.

Bayerische Feld - Luftschiffer - Abteilung 8

Zu dort. 2628 vom 3. 12. 16.

Dem Leutnant der Reserve H e p n e r spreche ich für sein schneidiges Verhalten bei dem Fliegerangriff am 9. 11. 16., welches zur Gefangennahme des feindlichen Fliegers führte, meine besondere Anerkennung aus.

Nach Abgang :

Der

Königlich Bayerischen Inspektion des Ingenieur-Korps

M ü n c h e n

vorstehende Abschrift unter Bezugnahme auf dort. Nr. 34684 vom 19. 11. 16. zur Kenntnis.

F. d. k. G. d. L.

Der Chef des Generalstabes

Thomsen

Oberstleutnant.

lines with extreme speed via Merville.

"Between the individual tethered balloons, at an altitude of 500 to 1000 meters, I continuously changed altitude and direction, curving sharply. In order to overcome the extraordinarily heavy machine gun fire as best as possible. As a result, the anti-aircraft fire was very limited.

"I only opened fire on each balloon at very close range. Flying horizontally at 1400 rpm, I flew close to each balloon until I rammed it and then pulled my Albatros just above the balloon. Almost all of them crashed to the ground covered in flames after 15 to 30 seconds. I attacked no balloon without success. By 4.30 a.m. all the balloons at the front had been retracted.

"All in all, I fired 100 phosphorus and 100 K cartridges. So on average I set fire to each balloon with 20 phosphorus and 20 K cartridges.

"I constantly distributed 400 phosphorus and 110 K cartridges for each gun so that the belt of the left machine gun had four phosphorus and one K bullet each. The belt of the right machine gun had one phosphorus bullet for every four K cartridges."

The most extraordinary balloon fighter is certainly 2Lt. Frank Luke, who shot down 14 German balloons and four airplanes in just 17 days. Born the son of German

Above: Albatros D.Vs of Jasta 15 at Le Clos Ferme airfield. Lt. Gontermann can be seen on the right side of the picture in front of his camouflaged Albatros.

Above: Lt. Gontermann in front of his Albatros with the red fuselage band and colored bands attached to the tailplane.

Left: After the five air victories of Lieutenant Friedrich Röth on May 29, 1918, Prof. Zeno Diemer painted this dramatic picture for the press. (Sophie Diemer).

Albatros D.V of Jasta 15, camouflaged for balloon hunting with gray-green spots on the fuselage and light spots on the wings.

Above: Frank Luke in front of his Spad XIII C1 No. 26 in Rembercourt. (Jack Herris)

Above: On September 13, 1918, Luke shot down an LVG C.VI of FA 36 as his 13th aerial victory. (Jack Herris)

Above: Above the village of Milly-sur-Bradon, Luke set course for Murvaux.

Above: 2Lt Luke during target practice with a German Maxim MG. (Jack Herris)

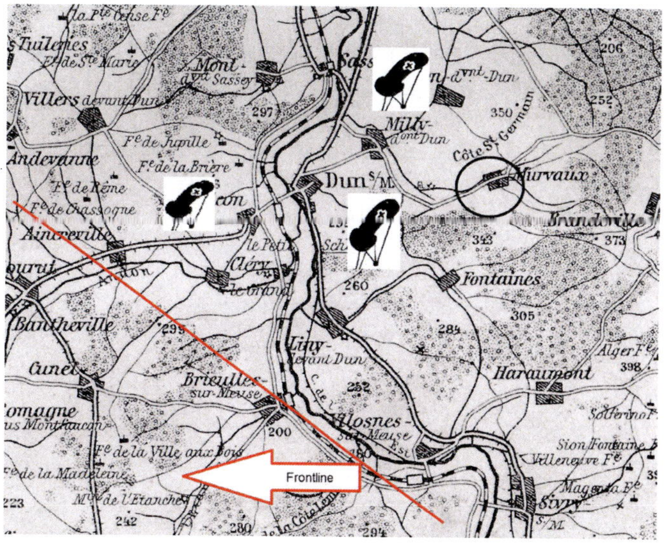

Above: Frank Luke's last flight on 29.September 1918.

immigrants on May 19, 1897 in Phoenix, Arizona, he volunteered after the USA declared war on Germany and joined the Signal Corps in September 1917. At the School of Military Aeronautics in Austin, Texas, Luke received pilot training and was assigned to the 27th Aero Squadron (First Pursuit Group) in France in March 1918. On September 12, with the 27th Aero Squadron now based at Rembercourt airfield with Spad XIII, Frank Luke began his career as a balloon fighter. He shot down a German balloon near Marieulles, which he immediately had confirmed by members of a US balloon company. By September 28, he had increased his aerial victories to 15 (11 balloons and 4 airplanes).

On the evening of September 29, Frank Luke took off towards the German front, where he shot down three balloons within a few minutes near Avocourt. These aerial victories were confirmed by the observers of the 7th Balloon Company, which was located near Souilly. Luke must have been seriously wounded by the flak during the attacks on the balloons. After he had shot down the third balloon, he certainly lost his orientation, because instead of flying towards his own lines, he set course for the German-occupied hinterland. We can only speculate as to whether Frank Luke lost consciousness part way through the flight or whether his vision was impaired due to severe blood loss. He landed with his Spad XIII in a meadow near Murvaux and then went to a stream that ran alongside the road. Probably 2Lt. Luke wanted to quench his thirst here (belly shots often evoke a feeling of thirst). Luke opened fire with his pistol at the German soldiers who had rushed over and were now also firing. Frank Luke was fatally wounded in the process. After Luke fell, a medic examined him and took his papers. These were later handed over to the American ambassador in Switzerland via the Red Cross in Geneva. It is not known why Luke was not wearing an identification tag; part of the tag usually remained with the body to enable later identification. Frank Luke could only be identified in his grave in the church cemetery in Murvaux in 1919 on the basis of his wristwatch. It is highly unlikely that German soldiers robbed him of his personal belongings. Body snatching was severely punished. Capt. George Guynemer's personal belongings (4 identity cards, letters and notes) were also taken by a medic of the 204th Infantry Division after his death in Poelcapelle. These were also handed over to France via the Red Cross in Geneva.

Frank Luke now rests in the American Cemetery Meuse-Argonne, near Romagne-sous-Montfaucon. Posthumously, the US Congress awarded 2nd Lieutenant Frank Luke the highest American decoration, the Medal of Honor.

Endnotes

1. Schwarte, M: *Die Technik im Weltkriege*, Berlin 1920.
2. Langsdorff v., Werner: *Flieger am Feind*, Gütersloh 1934.
3. Wir Kämpfer im Weltkrieg: Bericht des Offiziersstellvertreters Heibert.
4. Royal Flying Corps Communique No. 87.
5. Bayerisches Hauptstaatsarchiv-Abt. IV Kriegsarchiv: Akte ILuft 40.
6. Langsdorff v., Werner: *Flieger am Feind*, Gütersloh 1934.
7. Brzenk, Hans: *Im Kampf wider Kokarde und weissen Adler*, Berlin 1925.
8. Diese farbigen Bänder waren auch das Zeichen des Staffelführers.
9. Heydemarck, Georg: *Der Adler des weißen Meeres,* München 1921.
10. Guttman, Jon: *Ballon-Busting Aces of WW 1*, page 89 ff.
11. Coppens, Willy: *Jours Envolés*, Paris 1932.
12. Schmäling, Bruno / Bock, Winfried: *Royal Bavarian Jagdstaffel 23*, Reno 2018.
13. Wir Kämpfer im Weltkrieg: Bericht des Lt. d.R. Röth.
14. See also: Durkota, Alan E: *Medal of Honor Aviators of WW I*, Reno 2021.
15. Total losses on 29. September: 11 Ballons, dead: 1 Offizier and 5 Soldiers.

Spad 13 "26" flown by First Lt. Frank Luke, 27th Aero Squadron 18 total victories, 14 balloons Luke was the leading American balloon buster.

Above: Murvaux in the summer of 1917.

Above: The village of Murvaux today, with the church in the background.

Pfalz D.IIIa
Friedrich *Ritter* von Röth
***Jasta* 16b**
28 total victories, 20 balloons
Von Röth was the leading German balloon buster.

Albatros D.V
***Lt.* Heinrich Gontermann**
***Jasta* 15**
39 total victories, 17 balloons

Above: The village cemetery behind the church in Murvaux, where Frank Luke was first buried.

The Meuse-Argonne American Cemetery in 1920.

Above: A total of 14,246 fallen US soldiers rest in the cemetery.

Right: The grave of 2Lt. Frank Luke.

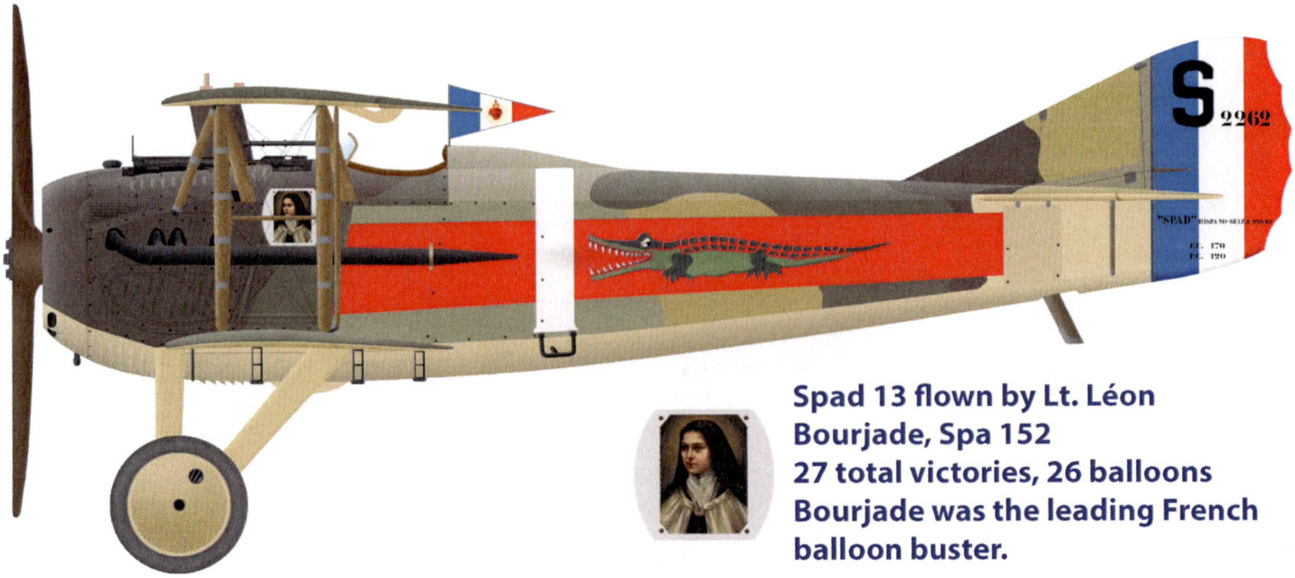

Spad 13 flown by Lt. Léon Bourjade, Spa 152
27 total victories, 26 balloons
Bourjade was the leading French balloon buster.

Naval Captive Balloon Department

Above: The balloon of the naval tethered balloon department near Steene before the ascent.

Above: The *Deutschland* battery with four 38 cm SK L/45 guns was stationed on the coast between Breedene and Klemskerke.

Above: A 38 cm L/5.4 explosive grenade weighing 750 kg with a firing range of 38,700 m.

Above: The 38 cm *"Pommern"* battery near Moere fired balloons far behind the Belgian front.

From spring of 1915, the Marinekorps Flanders utilized balloons to survey the occupied coast between Dutch border and Westende-Bad with its adjoining waters. For this, Kommandeur des Marineluftfahrwesens (K.d.L.F.) des Marinekorps, Kapitän z. S. Hans Herr, had the Marine-Fesselballon-Abteilung at his disposal which included one balloon and the Marine-Lichtbild-Trupp (naval photo troop). Using a Zeiss 120 cm camera, observers captured panoramic pictures, paying particular emphasis on the front section on the river Yser. Their work was considerably more complicated because of mist above the lake and river Yser area. Photographers had to wait for the few clear visibility moments to capture their images, and despite the use of yellow filters, achieving a clear image capture was rare.

With the reinforcement of the coastal area by naval guns, the tasks of balloon observers expanded to include naval artillery fire control. From mid 1915, the navy moved heavy coastal batteries to the Flemish coast in the area of the I. Marine-Division. From late September 1915, the *Tirpitz* battery was ready to fire. It consisted of four 28 cm SK L/45 cannons reaching up to 34 000 m. By the end of 1916, 38 cm batteries followed in *Pomerania* near Moere and in *Germany* near Breedene.

Additional navy captive balloons followed in February 1916; among other areas, they were tasked with surveying the Dutch coastal areas and the Scheldt estuary. There, the army and Marine-Feldluftschiffer worked closely together. Half of the division led the army (Hptm. d.R. Herr) near Uytkerke while the other part of the division near Steene was led by the navy (Lt.d.R. MA Schultz).[1]

Initially, the navy only had MG and carbine rifles with a 25-round magazine capacity to defend against enemy air attacks. All balloons were equipped with parachutes which were treated with a fireproof impregnation agent on July 20, 1916. In September of 1916, the Flanders corps was assigned an additional Ballonzug stationed in Morin near Zevecote.

Lone balloons on the Flemish coast were often targeted by enemy fighter aircraft. On September 7, 1916, Uffz. Hadeball used a 120 cm camera to capture the front along the river Yser from Steene. Around 16:00, Oblt. Forstreuter ordered for the 1000 m^3 balloon to be retracted due to hazy conditions. Uffz. Hadeball was looking forward to a cup of warm coffee after spending several hours in the basket of the balloon when he spotted an enemy aircraft at an altitude of 500 meters.

On the ground, aircraft observer Maat Lorenz also noticed the aircraft, but the balloon quickly caught on fire. Eight Le Prieur rockets had been fired at the balloon from the enemy aircraft. Unteroffizier Hadeball immediately

Above: The balloon hangar of the naval tethered balloon department north-west of Steene.

Above: Locations of the naval tethered balloons on the Flemish coast.

jumped out of the balloon when he heard the hissing of the approaching rockets. As the balloon was pulled down at rapid speed, the parachute, which opened late, quickly moved away from the balloon. For several seconds, the balloon remained in the air before it toppled to the ground, passing the parachute. The observer landed in a meadow approximately 70m from the point of explosion of the balloon, unscathed.

On October 20, 1916, the navy balloon near Steene was once again targeted by an enemy aircraft. RNAS Lt EW Norton attacked Steene's half division 1000 m^3 balloon east of Wilskerke at approximately 12:30. The balloon was at an altitude of 650 meters with the base of the cloud ceiling at 800 meters. Lt Norton had approached under cover of the clouds, firing Le Prieur rockets at the balloon. When observer Lt. Bendixsohn jumped out of the balloon basket, his parachute opened correctly. Despite engaging the motorized winch to move the balloon against the wind as soon as the aircraft was discovered, the burning balloon casing contacted the parachute, causing it to plunge to the ground and crash. The parachute partially opened, allowing Lt. Bendixsohn to land simultaneously with the balloon in a field east of Wilskerke. He sustained slight injuries to the back of his head as well as a fracture of his left upper thigh. Fireproofing of the parachute saved Lt. Bendixsohn's life; the parachute remained intact save a few burn holes.

In early 1917, the Marine-Fesselballon-Abteilung consisted of three troops stationed in the vicinity of Knocke, Steene, and Zevecote. The division's manual winches were equipped with 80 horsepower Argus engines in February; in March, Feldluftschiffer-Park 2 in Maubeuge delivered new Ae balloons in a grey-green color.

On April 29, 1917, naval balloon divisions were subordinated to the army. Initially, the term Marine-Fesselballon-Abteilung was preserved as it had originally been established as part of the Reichs-Marine-Amt's (RMA) now decommissioned special purpose vessel, "*Indianola*". In the summer of 1917, FLA 3 deployed to Wenduyne to support the Marine-Fesselballon-Abteilung. British and French soldiers repeatedly attempted to break through the front lines, particularly on Yser estuary.

Above: A comrade of the Seefrosta from Nieuwmunster protects the naval balloon near Steene from enemy fighter planes.

Marine Corps aircraft were not able to offer much protection to the balloons; for this reason, heavy anti-aircraft guns were involved, including 7.7 cm L/35, 8.8 cm L/45, as well as 10.5 cm L/45. On June 27, 1917, enemy aircraft attacked balloons 2 and 3 while they observed 38 cm batteries, "*Pommern*" and *"Deutschland";* the attack was warded off by anti-aircraft guns.

On October 24, 1917, the Marine-Fesselballon-Abteilung was dissolved; it was transferred to the army, renamed FLA 47. As a result, the reconstruction of the steamboat, "*Rio Pardo*" into a balloon ship in Zeebrugge was cancelled. The vessel was to direct coastal artillery from the sea.

Endnotes

1. NARA: T1022.

Above: Army and Navy airmen in the service briefing.

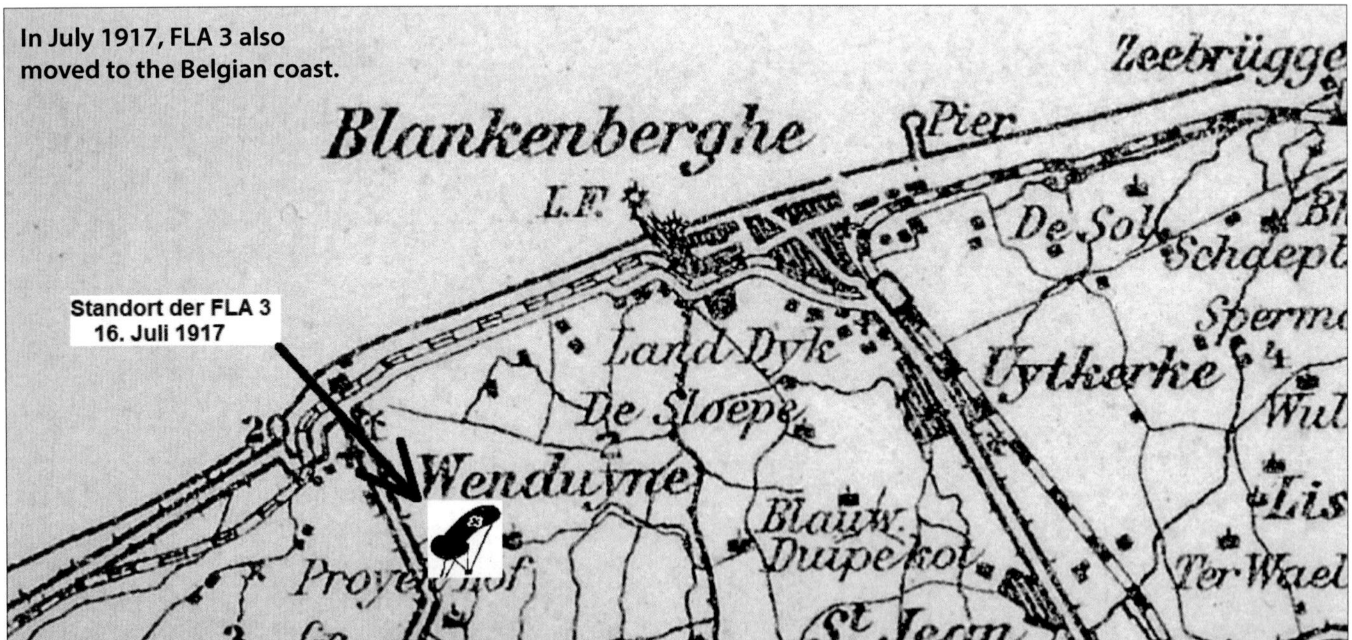

In July 1917, FLA 3 also moved to the Belgian coast.

The k.u.k. Balloon Division

Above: The parachute of the K.u.K observer is housed in the container on the right side of the basket. (NARA)

The allied Austrian-Hungarian Luftfahrtruppe established 27 balloon divisions through the end of the war. The Luftschiffer-Schul-Kompanie established in Traiskirchen on March 21, 1917, provided replacements and training. In late summer of 1917, the balloon division was designated Ballon-Kompanie; it consisted of six officers, 128 men, one balloon, one gas car, one motorized winch, four horse drawn vehicles, and 28 horses. From 1917, K.u.K. Ballon-Kompanien introduced the Ae balloon type. *Brunner* type parachutes were part of the standard balloon observer equipment from 1917. Until the end of the war, 169 parachute jumps took place, of which two ended fatally.

On November 30, 1917, Ballon-Kompanie 13 Lt. Alfred Bibicas jumped out of his balloon too late; the fire of the burning balloon set ablaze his parachute. Ballon-Kompanie 20 Lt. Freiherr von Scholten fell to his death on May 8, 1918, after his parachute failed to open.[1]

Above: Transportation of the balloon to the ascent site on the Galician front; the telephone van can be seen in the background. (NARA)

Above: Ascent of the balloon. (NARA)

Above: German comrades visiting an Austro-Hungarian balloon company on the Eastern Front in 1917 (NA-RA).

Left: The motorized winch was driven by an Austro-Daimler engine. (NARA)

Below: The K.u.K. Balloon companies continued to use the Parseval-Sigsfeld type until the end of the war. (NARA)

Balloontruppe Personnel

Above: Lt. Thofehrn, left, with comrades in the quarters.

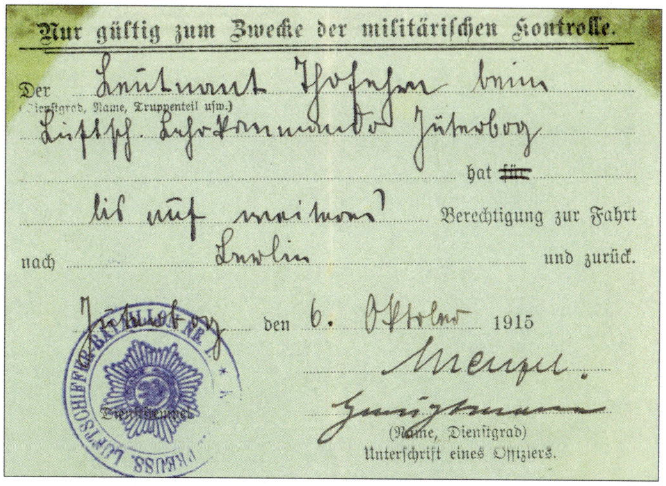

Nur gültig zum Zwecke der militärischen Kontrolle.

Der Leutnant Thofehrn beim
(Dienstgrad, Name, Truppenteil usw.)
Luftsch. Lehrkommando Jüterbog
hat
bis auf weiteres Berechtigung zur Fahrt
nach Berlin und zurück.

Jüterbog den 6. Oktober 1915

Menzel.
Hauptmann
(Name, Dienstgrad)
Unterschrift eines Offiziers.

Above: During his training in Jüterbog, Lt. Thofehrn was allowed to visit Berlin after his service.

Feldluftschiffer-Abteilung personnel originated from all different military branches. Artillery officers were particularly well-suited as observers because they possessed the necessary skill and knowledge for this task. They received additional training at the Luftschiffer-Schule to further their abilities in ground observation, photography, and the shooting in of all artillery calibers. An experienced observer was present in the balloon basket during the first three to five months of training. Thereafter, observer students who were not part of an artillery troop were ordered to join the heavy artillery on the front for a few weeks. There, they gained the necessary experience directly from the artillery. Upon completion of their time on the front, a two-month long training period followed in which students learned to zero in artillery at the Luftschiffer-Schule. At the artillery training grounds, observers then implemented what they learned in cooperation with all types of artillery gun calibers. Once observers mastered these tasks, they were transferred to the front in the role of assistant observers. There, under the supervision of experienced observers, they completed their training; special emphasis was placed on the collaboration with Ballonzentralen and Staff. Photography training was provided by the Schulen der Inspektion des Lichtbildwesens (schools of the inspection of photography). The processing

Above: The training also included taking photos from airplanes. The aerial photo shows the Victory Column and the Reichstag in Berlin.

Above: Balloon trains and the artillery of FAR 22 practiced working together at the artillery training area in Signy.

Above: Lt. Thofehrn photographed the Marienburg during an exercise in the fall of 1915.

Above: HQ of the VIII Reserve Corps, special observations had to be reported directly to the staff. Lt. Thofehrn also reported here with the photos from the balloon about the situation on the other front.

of photographic optical discs, as well as the development of discs and the production of photographic prints was carried out by ground personnel.[2] Ground personnel were also trained at the Luftschiffer-Schule; most of the candidates were mechanics, locksmiths, electricians, or photographers.

Balloon observers needed to be physically and emotionally resilient; oftentimes, duty in the unsteady balloon basket could extend ten hours or more. Even experienced flyers who were guests in balloon baskets ended up airsick within a short period of time. On November 6, 1917, the 4th army's FLA 19 and 36 observers (Bz 2, 3, 6, and 92) stayed in their balloons for hours in driving rain, cold temperatures, and winds up 20 meters/second, directing their own artillery during a British attack. Bz 124 Lt. Reimers ascended to 1000 meters. Tasks such as dealing with balloon tether breakage required observers to have nerves of steel. On November 24, 1917, Bz 88 Lt. d.R. Tamm observed the enemy lines in the Alsace region. At approximately 16:00, the balloon tether tore, and the balloon drifted westbound, toward the city of Basle in Switzerland, ascending to an altitude of 3500 meters. Later, the balloon was caught in an air current carrying it towards Bad Säckingen in the Black Forest, where Tamm pulled the rip panel. Tamm reported back with his unit on the following day. On December 22, 1917, Bz 124 Lt. d.R. Kienzle was struck by the same fate. During a storm, his balloon's tether broke, and he drifted towards France. Prior to reaching the front, Kienzle tore open the rip panel and jumped out of the balloon in his parachute; he landed in Wervik. During the night, infantrymen recovered his balloon from no man's land.

Kurt Stollwerk is known as the youngest German Feldluftschiffer.[3] At age 14, he volunteered to sign up with Luftschiffer-Ersatz-Abteilung 3 in Cologne. In January 1915, he arrived at the front and joined the Luftschiffer-Bataillon led by his father. After a few months of duty on the front it became apparent that Stollwerk had not yet completed any military training. Since he was an experienced

Above: Lt. Thofehrn with fellow observers at Easter 1917.

Above: The gas column moves up to fill the balloon.

Above: The balloon is being filled. The inflation hose to the balloon can be seen in the center.

soldier, sending him to recruit initial training seemed inappropriate. He was therefore ordered to complete training with Reserve-Infanterie-Regiment 215 upon completion of which he returned to the Feldluftschiffer. Kurt Stollwerk carried out more than 300 balloon ascents; he was shot down ten times. On May 20, 1918, Vzfw. Stollwerk jumped out of a balloon twice with Bz 147. At age 17, Stollwerk was promoted to Leutnant d. Reserve in the summer of 1918.

The highest Prussian medal, Pour la Merite, and the Prussian Golden Cross of Merit (Preußisches Goldenes Militär-Verdienstkreuz) was only awarded to one respective member of the Feldluftschiffer-Truppe during WWI. On August 30, 1918, Offiziers-Stellvertreter Max Reiß was awarded the Preußisches Goldenes Militär-Verdienstkreuz,

Above: Lt. Thofehrn ascends with a second observer.

Above: The basket is being prepared for the ascent. Only one observer could ascend in this smaller basket. The map board is folded up. The bag for the Paulus parachute is attached to the outside of the bas-ket.

award number 1100. Reiß was part of Ballonzug 91.

Lt. Peter Rieper, born on April 13, 1887, grew up in Wisch near Stade. Upon completing his school education, Rieper studied chemistry in Jena. He wrote his thesis in 1912 and graduated Dr.phil. He completed his military service with Torgauer Feldartillerie-Regiment Nr. 74, when he left in 1913, his rank was Vizewachtmeister d.R. When the war

Above: A photo of the ground team from the basket.

Above: Officers of the Reserve Division observe the rise of Lt. Thofehrn.

Right: Lt. Thofehrn regularly visited the front line. On the one hand, he was able to hand over the latest aerial photographs immediately and, on the other, he received the reconnaissance requests of the infantry and artillerymen directly.

Below: The commander of Bz 45 Rittmeister Stach, on the right at the basket, and Lt. v. Holtzheim, left, at the Somme.

Above: Lt. Delius and Lt. Kruse in the basket. Lt. Delius crashed fatally near Courtelles on July 5, 1916, but is not mentioned in the casualty lists.

Above: Airman Karl Korell fell on May 15, 1916 near Niedermorschweiler in the Vosges mountains. The wrong date of death is indicated on his grave cross as May 5, 1916.

The observer of Bz 50, Lt. Stein, was awarded the Iron Cross 2nd Class in 1916

Lt. Peter Rieper was the only balloon observer to be awarded the Pour le Merite.

Casualties of Prussian Balloon Personnel		
Year	**Killed**	**Missing of POW**
1914	26	12
1915	90	–
1916	124	24
1917	132	4
1918	347	8

broke out, Rieper deployed to the western front with the 1. Batterie des Torgauer Feldartillerie-Regiments Nr. 74 where he was badly wounded near Mons on August 24, 1914.

Following the convalescence, Rieper returned to his regiment in November 1914 where he was promoted to Leutnant d.R. on December 24. In the summer of 1915, Lt. Rieper transferred to the Feldluftschiffer and was trained to become an artillery observer at Jüterbog. He was then assigned to Feldluftschiffer-Abteilung 7, and later Feldluftschiffer-Abteilung 21. Lt. Rieper was likely assigned with the leadership of Ballzug 19 in March of 1917. Rieper ascended in a balloon almost daily, delivering valuable information to the artillerists attached to Ballonzug 19. His observations were of great value during the Kaiserschlacht ("Kaiser's Battle" or German spring offensive) from March 21, 1918, with the 18th army and in May with the 7th army on the Chemin-des-Dames.

In the early morning hours of June 3, 1918, Lt. Rieper ascended in his balloon as usual. He was focused on his task of zeroing in his artillery on identified targets and did not notice the approaching Spa 81 Spad VII. The three Spad aircraft, flown by Adjudants Paul Cardon, Malfanti, and Paul Guérin,[4] had already begun shooting at the balloon when Rieper was ordered to jump out. Rieper was shot in the right arm while still in the balloon basket; additionally, he broke his fibula on landing. During his stay at Reserve-Lazarett 3 military hospital in Hamburg-Altona, Artillerie-Kommandeur Nr. 37 recommended him as a Pour-le-Merite recipient with Kogenluft. On July 9, 1918, Lt. Rieper received a telegram from Kogenluft General v. Hoeppner informing him that he had been awarded the Pour-le-Merite medal by the emperor.

Based on his injuries, Lt. Rieper was not to return to the front. He continued his service as an instructor at the Luftschiffer-Beobacherschule in Namur from October 1918 through the end of the war. According to Neumann,[5] Ballontruppe losses on the front (not including Bavarian units) amounted to 709 soldiers. Effectively, the number of fallen men was likely significantly higher as not all fallen soldiers were reported and included in the statistics. Among them was Lt. Delius who fell on July 5, 1916, near Courtelles on the river Somme. In 1936,[6] assistant observer Schreiber described Lt. Delius as a successful observer who would often reach altitudes of 1800 meters sitting on a piece of board. It was Delius's idea to use a yellow "Friedensballon" (peace balloon) a few hundred meters away from the actual balloon to lure enemy aircraft. According to Schreiber, however, this idea did not work.

Albatros D.V 2194/17
***Hptm.* Adolf *Ritter* von Tutschek**
***Jasta* 12 / *JG* 2**
27 total victories, 3 balloons

Balloons

Above: A Parseval-Sigsfeld balloon before its ascent in 1915.

As previously mentioned, the German Ballontruppe went to war in August 1914 utilizing the 600 m^3 Parseval-Sigsfeld Drachenballon (kite balloon). The balloon casing was made from double-layered, diagonal cotton bands; in between was a rubber sealing agent. Balloon fabric was strong enough to withstand rain, storms, and other environmental influences. The outer layer was dyed yellow to protect it from the sun's UV rays. Throughout the war, the balloon gas bag was camouflaged using green or black color; at times, multiple colors were used. The lack in raw materials forced balloon plants to utilize silk or silk blends in the balloon

Above: A last photo with the teams before the ascent.

Above & Below: Two-color camouflaged Ae balloon of Bz 45 near Orfenil. The balloons could also ascend with two baskets up to an altitude of 1000 meters.

manufacturing process from 1917. The caoutchouc sealing agent was replaced by synthetic rubber or Goldbeater's skin (the processed outer membrane of animal intestines).

The gas bag interior consisted of two chambers, one containing gas and one for air, otherwise known as a ballonet. The airtight ballonet wall was almost entirely pressed against the back part of the balloon's outer cover when the balloon was filled completely. When the gas chamber of the balloon was filled, excess air was released from the ballonet via a valve connected to the rudder lobe. A chain connected to the ballonet wall automatically activated the valve if the internal pressure of the balloon exceeded a predetermined limit. High internal pressure was caused by solar irradiation heating raising the gas temperature, or because of lower air pressure with an increase in altitude.

With a decrease in temperature or gas loss, the size of the gas chamber was reduced. For the balloon to maintain its full shape, air was forced into the ballonet. The mixing of air and gas was prevented by separating gas and air chambers within the balloon. The rudder lobe had an opening towards the front allowing the balloon to continuously be steered in the

Above: The Ae balloon rises in cloudy weather; the Paulus parachute is attached to the front basket.

direction of the wind. Steering was supported by two sails attached at the back part of the balloon.

In case of an emergency, observers pulled a line connected to the valve to open the gas chamber. Alternatively, they could utilize the rip panel. The rip panel was activated in case the balloon was uncontrollable; the observer then jumped out of the balloon basket with his parachute.[7] On the front, ground personnel were able to repair a rip panel within one hour.

To check a balloon gas bag, a blower was used to fill it with air; the responsible Offizier crawled into the balloon through the filling hole. Care was to be taken not to damage the inner casing. Felt slippers were worn instead of boots. Incident light indicated an untight area in the casing which had to be pasted over. The overpressure valve as well as the

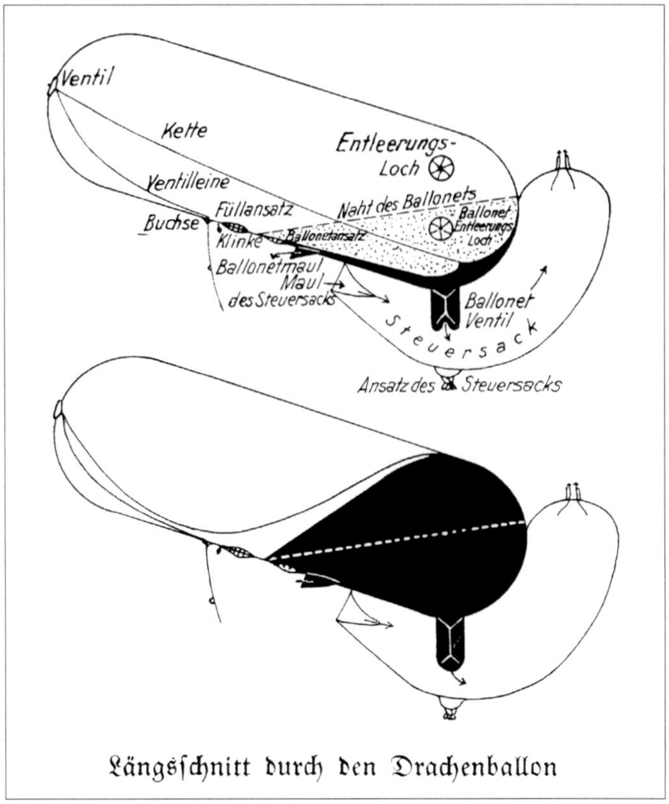

Above & Below: The construction of the Parseval-Sigsfeld balloon.

ballonet wall were visually checked by the Offizier.[8]

By 1915, ascent altitudes of 500-600 meters were no longer sufficient. To rise higher, observers experimented with lighter harnesses as opposed to balloon baskets; however, using the alternative device made it impossible to stay aloft for extended periods of time.

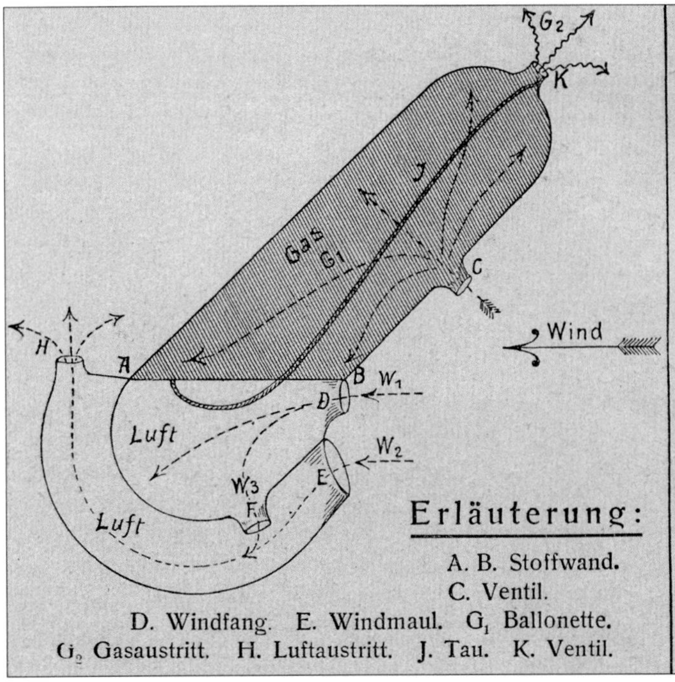

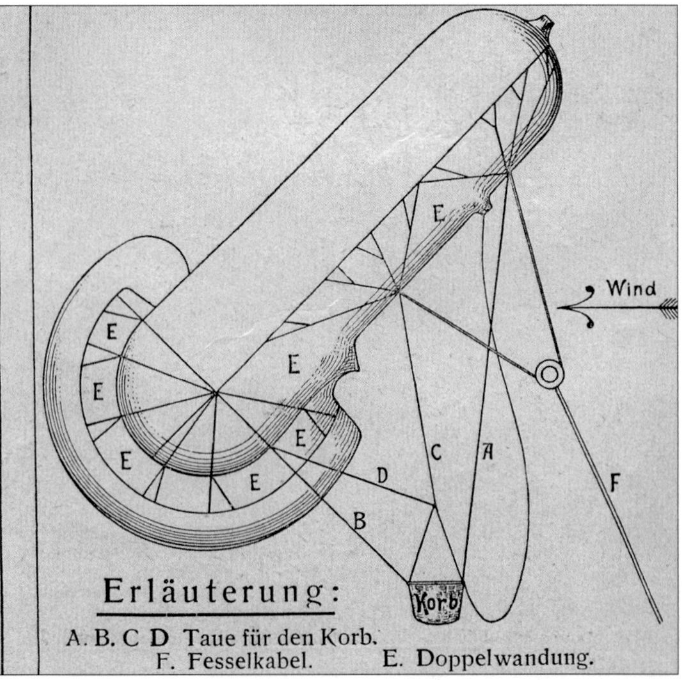

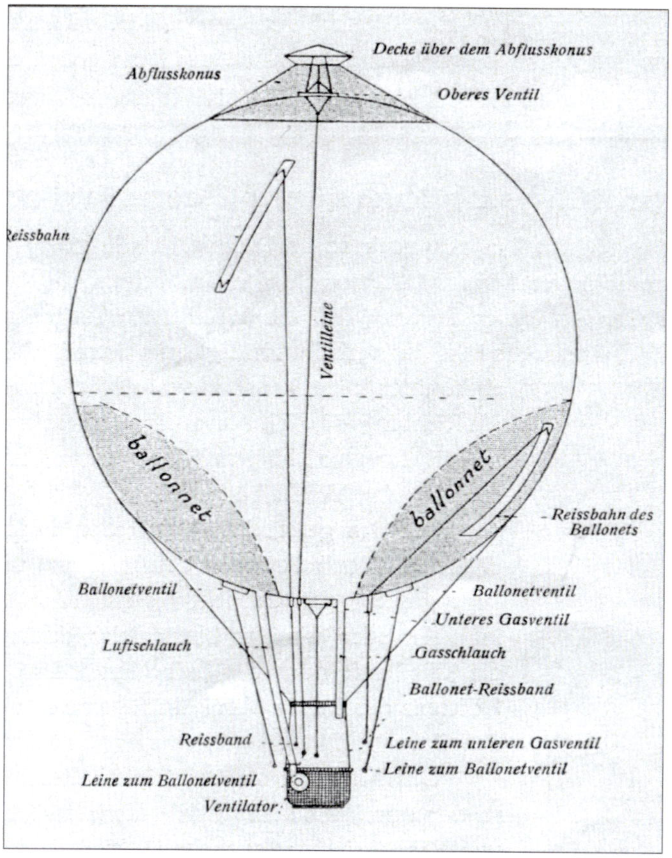

Above: Colored photo of a balloon showing the yellow linen color of the envelope.

Left: Drawing of a balloon.

Balloon volume capacity was therefore increased from 600/650 m^3 to 800/850 and 1000 m^3. Raw material supply became an increasingly more worrisome task. From 1916, cotton delivery for balloon bag production came to a halt. Silk and substitute fabrics were used instead. Caoutchouc sealing agents was increasingly replaced by synthetic rubber and Goldbeater's skin in 1917. By the end of 1916, caoutchouc was exported to Germany by the United States; the trade submarine *Deutschland* alone delivered 726 tons.

At times, Ballonfabrik Riedinger utilized varnish to repair

Above: At ground level, the balloon remained clearly visible despite the yellow paint, and it was only at altitude that the outline became blurred due to the color.

Above & Above Right: Preparing the balloon envelope for filling with gas. The balloon fabric is printed with a camouflage pattern. (NARA)

damaged inner balloon casings. This process did not prove to be successful, as dried varnish did not fully seal the casing. On August 9, 1916, a fatal accident occurred with a repaired balloon:

"Half division Berey balloon exploded. The balloon was assigned to 10. Reserve-Divison. Reason for loss was poor casing materials. The casing had been reprocessed by Fa. Riedinger, the inside casing had been coated with varnish.

During filling, the balloon began showing signs of deflation after the 193rd bottle. It did not rise higher than 800 meters, and at 15:45 it exploded in the air.

Vzfw. d.R. Josef Schmidbauer dead, Soldat Minderlein gravely injured."

When transporting a balloon gas bag, it had to be rolled up crease-free in accordance with manufacturer's regulations and then stored on the equipment truck. Manufacturers recommended storing balloons in balloon hangars during poor weather conditions. Balloon companies also offered hangars for purchase.

On June 4, 1916, Major Thomsen organized a supply of balloons and winches for the Feldluftschiffer-Truppe. The Prussian and Bavarian Inspektion (Idluft and Iluk) divisions were to receive priority in balloon and winch deliveries, particularly in the areas that were most heavily embattled.

FLA were required to report losses of balloons or winches to Berlin or Munich. All balloons were assigned a clear designation and number which was to be painted three meters above the balloon's filling base. The Feldflugchef assigned balloon numbers 1-6 to FLA 1, balloon numbers 7-12 to FLA 2, etc. to the Prussian FLA. The Bavarian numbering began at 101.[9]

Above: The Ae balloon takes shape, with the overpressure gas valve visible at the front.

Above: The windsocks on the balloon served to stabilize the Parseval-Sigsfeld balloon.

Above: The tear line on the balloon envelope is clearly visible here.

In the summer of 1916, a new balloon type, the Caquot Type M, surfaced in Flanders. From the fall, British and Belgian Ballontruppen also began utilizing this balloon type. German observers noticed that this balloon type was considerably more stable due to its three rudder lobes. Furthermore, it showed a near horizontal position while

Above: In an emergency, the observer could use the lines to open the valve or the tear track in the gas compartment.

Above: The effect of the control bag can be seen on this anchored balloon. The balloon automatically turns in the direction of the wind.

capable of ascending to heights of 1500 meters. Neither observations with binoculars nor Flieger-Abteilungen supplied photographs made it possible to determine the construction of the balloon. It was not until a British balloon landed on the German side undamaged that the secret was revealed. The Type M balloon was developed by Albert Caquot in Chalais-Meudon; it was characterized by

Above: Inserting the gas valve into the balloon envelope at the airship school in Berlin-Reinickendorf before the war.

Above: The overpressure gas valve at the nose of the balloon.
Below: To check the inner balloon, it was filled with air. Here, the teams fill the envelope with a hand pump.

Above: The map board was attached to the outside of the balloon to save space.

its aerodynamic teardrop shape with three, interconnected rudder lobes.

The Inspektion der Luftschiffer-Truppe (Iluft) immediately had the captured Caquot balloon transported to Berlin where German and Austrian balloon manufacturers conducted a thorough examination. Soon after, a copy of the Caquot was developed, referred to as Type Ae (A=800, e=English).

The French origins of the Caquot was disregarded as the captured balloon was delivered by the Britsh Kite-Balloon-Service RFC.

Ballon-Hüllengesellschaft mbH[10] in Berlin-Tempelhof, Continental-Cauotchouc- & Gutta – Perca Comp. in Hannover, Luftfahrzeug-Gesellschaft in Bitterfeld, August Riedinger AG in Augsburg, and Vereinigte Gummiwarenfabriken in Harburg and Vienna produced a total of 1870 balloons. The cost of one balloon was 21,000 Deutsche Mark.

August Riedinger Ballonfabrik Augsburg AG

On April 1, 1897, August Riedinger and Johannes Scherle

Above: The tight space in the basket meant that the map board had to be attached outside.

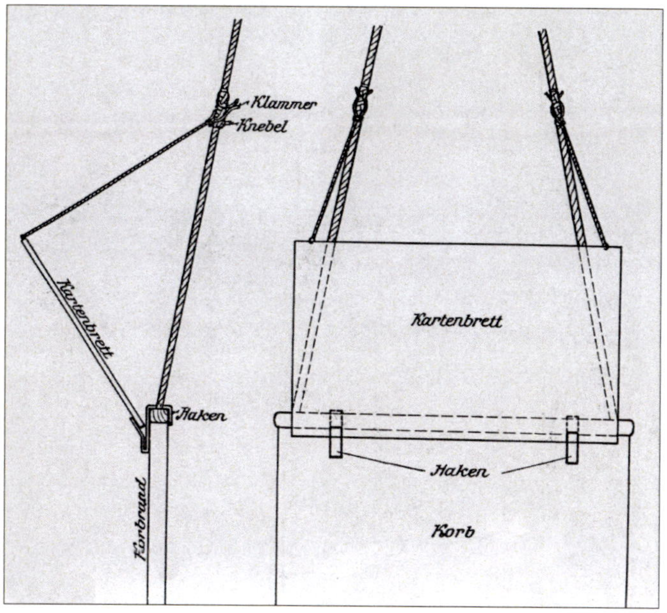

Above: Observers from FLA 20 (Bz 112) constructed a simple device that allowed the map board to be easily dropped in an emergency.

founded the Ballonfabrik, Augsburg. Riedinger had developed an interest in aviation in his youth and began to design flying machines and instruments. After two years of work, Riedinger finished the construction of his flying machine in 1892; however, a suitable engine was still missing. For this reason, Riedinger was forced to abandon his project and directed his attention toward balloon production. Riedinger was able to solicit the assistance of Hauptmann von Sigsfeld and Oberleutnant von Parseval.

Following the first successful Sigsfeld – Parseval kite balloon ascent, the patent licensing took place on October 10, 1883. The kite balloon underwent a three-year long test with the German army before it was introduced to the Luftschiffer-Truppe. In 1907, the Ballonfabrik was transformed into a GmbH (Ltd, limited company) which simultaneously took up production of gas cells and gas bags

Above: The Zeiss L30 camera with image plate.

Above: Zeiss L70 camera attached to the lines.

for Schütte-Lanz and Parseval airships. In 1913, 38 kite balloons were produced. According to a 1922 production brochure, available volume options ranged from 650 – 1000 m³. The August Riedinger Ballonfabrik Augsburg AG was in existence since 1915. The number of employees increased

Above: All the ropes were attached to the circumferential belt.

to 800, most of the personnel was female. In addition to balloons, Riedinger produced spherical balloons designed for weather observation, aircraft camouflage materials, parachutes, and balloon winches. Riedinger purchased the winches from other companies, one of which was Maschinenfabrik Keller & Knappich, Augsburg. The ceasefire ended the balloon production, forcing Riedinger to lay off most of his employees. Following WWI, the company manufactured high quality furniture until the production of free air balloons was permitted after the Versailles Treaty was lifted. Ballonfabrik Augsburg closed its doors in 2008.[11]

Through 1916, balloon ropes made from hemp were delivered by Italian manufacturers. After Italy officially entered the war, ropes were produced domestically using cotton or linen. Ropes were attached to a strap sewn to the balloon's outer cover. They were color coded as follows: red=

Above: Both balloon types remained in service until the end of the war. In the foreground the Caquot balloon type Ae, behind it the Parseval-Sigsfeld type.

Above: The Riedinger balloon factory was one of the best-known manufacturers.

Above: The company founder August Riedinger.

Above: After August Riedinger's death in 1919, his employee Johannes Scherle continued to run the company.

Riedinger gained important experience by building free balloons.

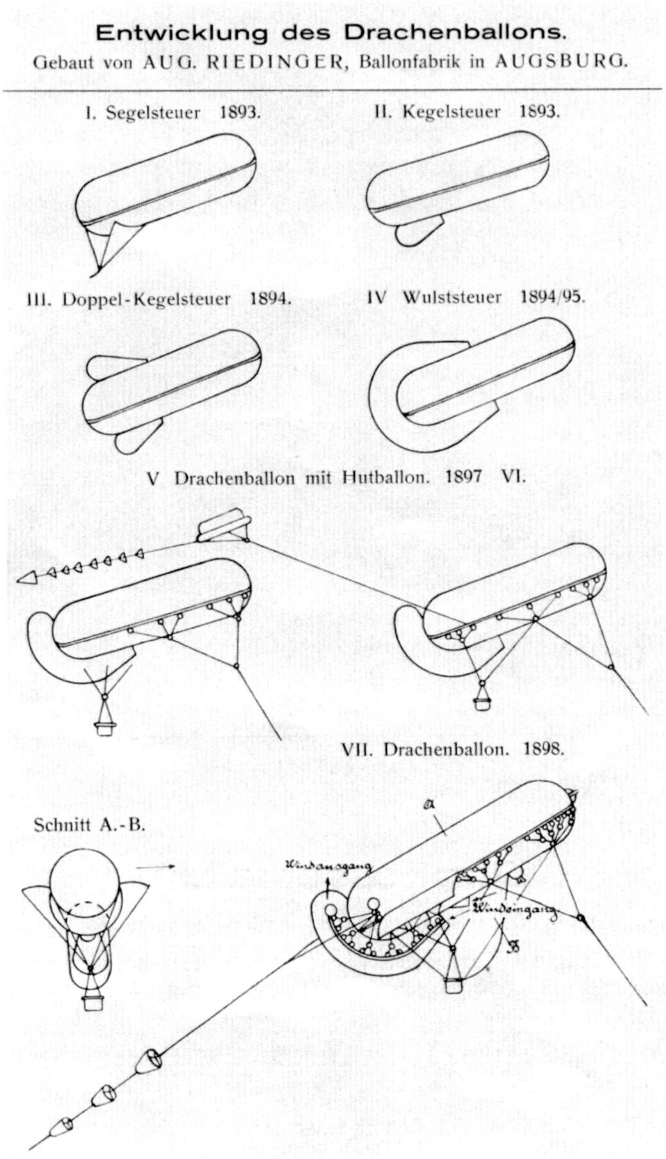

Above: The hydrogen gas was able to escape within a short time as a result.
Right: Overview of the pre-war balloons of the Augsburg balloon factory.

basket suspension, white = retaining rope, and blue = steering lobe rope.

The steel cable connecting the balloon to the winch measured 8-15mm, depending on wind speed. As this cable was the direct connection between the balloon and the ground, it had to be sufficiently insulated to prevent current voltage causing spark formation. The balloon winch deflected potential lightning strikes; from there, lightning was deflected into the ground by way of a grounding cable.

For safety reasons, orders were issued for immediate balloon retraction when thunderstorms were imminent. On June 28, 1917, the Bz 110 Ballonzugführer ordered the balloon to be retracted around 2000 as a thunderstorm was building. When the balloon was down to a height of 50 meters, suddenly, heavy enemy artillery fire began.

Despite the associated danger, observer Lt. d.R. Moeller

Above: Seamstresses in the production hall of the Augsburg balloon factory.

Right: Riedinger ventilator for filling the balloon envelope with air.

Above: Riedinger transport wagon for the balloon envelope.

Above: The balloon envelope was not allowed to lie directly on the ground during storage. To protect the envelope from moisture and sharp stones, the soldiers of FLA 2 built a wooden scaffold for safe storage.

Above: The gas cylinders could only be stored directly next to the balloon in a protected balloon hangar.

ordered for the balloon to ascend again. During the following hour, Moeller reported 78 enemy batteries when lightning struck. The electrical charge did not cause the balloon to catch on fire but Lt. d.R. Moeller was struck by lightning which caused temporary paralysis of his left arm for several hours. On July 19, Moeller received a personal letter of thanks from Generalleutnant Ernst von Hoeppner, commending him for his courage. Bz 53 was not as lucky on October 15, 1917. Despite army meteorological observatory level 3's (3rd army) thunderstorm warning, Bz 52, 53, and 54 balloons remained airborne for surveillance work. At 15:30, Bz 53 balloon No. Ae 857 was struck by lightning at an altitude of 850 meters. Lt. d.R. Graß and Gefreiter Menne were in the balloon basket and did not notice the lightning strike. Neither did the crew operating the winch, but telephone operators noticed strong electric shocks. The balloon was pulled down immediately; at an altitude of 500 meters, it suddenly flared up. Graß and Menne jumped out of the balloon right away and landed safely despite the burning balloon bag contacting both of their parachutes, multiple times. The neighboring Bz 52 and 54 balloons were also struck by lightning; neither of them caught on fire. A French balloon which was in the vicinity of La Neuville on the western edge of the Argonne was also struck by lightning; it caught fire and crashed.

The balloon personnel received sufficient instruction in the handling of highly flammable hydrogen gas. Nevertheless, negligence continued to be the reason for multiple accidents. On May 3, 1917, FLA 35 reported a serious accident:[12]

"Stoluft - Heeresgruppe Herzog Albrecht von Württemberg

I am reporting the partial burning of balloon Ae 606 H on May 3, 1917, in an oxyhydrogen explosion. The balloon did not hold any gas, apparently because of a rip panel leakage. The balloon had ascended in the morning; it was anchored and had lost approximately half of its gas contents

Above: Erection of a Riedinger balloon hall in Berlin-Reinickendorf.

Above: Balloon hangar built by the airmen themselves at the front. (NARA)

Above: Camouflaged balloon hangar at the front.

Above: The Bavarian coat of arms on the hall at Merchem could not be missing from the Bavarian FLA 8.

in strong solar irradiation. (…) As the loose outer casing was moving considerably in the wind, I ordered Lt. Witzig to empty the balloon shortly after 18:00.

I was nearby when I received notice that the balloon had exploded during the discharging process. Lt. Witzig and two men sustained grave burning injuries.

I immediately questioned everyone and came to the following conclusion:

The balloon was almost empty. Lt. Witzig ordered the

Balloon Markings from June 4, 1916

Marking	Capacity	Nr.	Company
B.T 1 H	1000 cbm	1	Vereinigte-Gummiwarenfabriken, Harburg und Wien
B.T 2 R	1000 cbm	2	Riedinger Augsburg
B. A 3 H	800 cbm	3	Vereinigte-Gummiwarenfabriken, Harburg und Wien
B. S 4 R	600 cbm	4	Riedinger Augsburg
B. 5 F.	750		Captured French Balloon
B. 6 B.	750		Captured Belgian Balloon
B. 7 E.	750		Captured English Balloon

Above: Balloon hangar of FLA 8 near Wytschaete in Flanders.

Above: The balloon of FLA 208 being pulled out of the hall and filled with gas again.

Above: In strong winds, the balloon observers could also ascend with kites. However, this did not prove successful as the observer had to sit motionless for a long time and there was hardly any room for maps. (NARA)

slightly flapping hose attachment to be rolled up at the deflation point; in the process, he gained a glimpse inside the wide-open deflation cavity. At that exact time, a large flame shot out of the deflation opening with a loud bang, considerably burning the faces and hands of the two men standing near it. Lt. Witzig sustained bad burning wounds to his face. His eyes remained unharmed; the glasses must have protected them.

German Balloon Losses 1914–1918

Year	1914/1915	1916	1917	1918
Enemy Planes	3	37	116	315
Enemy Artillery	10	8	20	37
Accidents	17	11	32	49
Total	30	56	168	401

Above: A Parseval-Sigsfeld balloon after being winched in with the hand winch. The 450 kg winch on the carriage was pulled by six horses.

Above: The hand winches built in 1896 remained in use until 1915.

Right: Horses remained indispensable to the airship crews until the end of the war.

Left: Transporting the hand winches on the protzenwagen caused major problems in rough terrain. (NARA)

Above: Thanks to the motorized Sedan winches, the balloons could be retracted much faster and also moved on roads with "balloon up". (NARA)

Above: The motorized winches were mostly offered by agricultural companies or machine factories, such as Alpine-Maschinenfabrik GmbH.

The fire in the balloon cavity was put out using sand; most of it was preserved."

Since 1915, FLA 35 was assigned to VIII. Reservekorps which in turn was part of the 3rd army since September 1914. There, FLA 35 partook in the Winter Campaign in the Champagne region. Through 1916, a static warfare took place between Ripont and Tahure.

For further protection, the German and allied forces balloon outer bags and parachutes were coated with a fireproof impregnation agent which helped lessen the effects of the incendiary ammunition. On September 4, 1918, Jasta 73 Lt. Franz Kirchfeld experienced this during a low-level attack on a French balloon:

"On 4.9.18 at 10:30 in the morning, I departed with a 4 aircraft Kette on a fighter flight west of Reims. As I did not encounter any enemy aircraft on the front, I descended onto two French balloons near Chambrecy, attacking from an altitude of 4000 m. I attacked the first balloon at 800 m from below using approximately 100 rounds at 150 m and closer. I saw the bullet in the balloon and watched the observer jump out, but the balloon

Left & Below: Riedinger also offered motorized winches in its company brochure.

Left: Retrieving the balloon with the motorized winch only took about 5–10 minutes.

Right: The Daimler power winch was also able to overcome difficult terrain thanks to the special iron wheels (gripper system). (NARA)

Winch Markings from June 4, 1916			
Marking	**Nr.**	**PS**	**Company**
B. 1. L. 35	**1**	**35**	**LFG**
B. 2. R. 32	**2**	**32**	**Riedinger Augsburg**

did not catch on fire."

Impregnating of parachutes with a fireproofing agents saved the life of a Bz 55 observer on July 7, 1918. Two Escadrille 314 Spad XIII aircraft attacked the balloon, setting it ablaze on their second approach at 300 meters: during the attack, Sgt. Fancois Castaings piloting Spad XIII (No. 4715) was hit by anti-aircraft guns and collided with the burning balloon, crashing fatally. The balloon observer's parachute was engulfed in the flames of the burning balloon but did not harm him. He was unscathed except for several light burning wounds to his face and hands.

The Balloon Basket

The frame of the balloon basket consisted of Spanish cane reeds, its walls of weaved willow reeds. Above it was a concentration ring made from steel or wood to which the balloon basket was attached with ropes. Standard equipment included a barograph as well as a barometer for altitude measurement, a statoscope to indicate climbs or descents, and a variometer to measure the rate of climb or descent. In addition, there was a compass, a pair of binoculars, maps, signaling equipment, as well as a camera with imaging discs. One or two Paulus parachutes were attached to the outside of the basket or the basket tether line. An equipment truck was used to transport the balloon bag, basket, and the ropes.

Balloon Winches

The Ballontruppe's winches were stored on a winch truck which was pulled by six horses along with the limber wagon. At the start of the war, the Truppe had 50 hand winches available. Using manual operation, down a balloon from 1000 meters took approximately one hour, too much time during an active threat posed by enemy aircraft or artillery. Luftschiffer therefore developed an alternative; they constructed motorized winches and Sedan winches. The workshop of the agricultural division in Sedan built clutches for 50 engines which were connected to the manual winches.

Manual winches brought in for repair at Inspektion der Luftschiffer-Truppe (Iluft) were equipped with engines (80-100 HP) under the direction of Major v. Kehler. As a result, the pull-down speed from an altitude of 1500 meters was reduced to five minutes in medium wind conditions. The motorized winches, each weighing 4700 kg, remained mounted to the horse drawn trucks. From June 4, 1916, motorized winches were issued clear markings and numbering, for example: the Prussian FLA 1 winch numbers 1-6, FLA 2 winch numbers 7-12, etc. The Bavarian numbering started at 101.

Iluft purchased a total of 330 motorized winches from a variety of manufacturers, priced at 21,000 Mark per piece. In 1918, industrial towing tractors equipped with motorized winches were introduced, allowing balloons to be moved across rough terrain. This was particularly important when the balloon was under fire by enemy artillery. Towing tractors made it possible to move balloons not only vertically but also horizontally.

Gas Wagons

Above: Transport trolley for 6 gas cylinders of the Bavarian FLA 8 on the Somme.

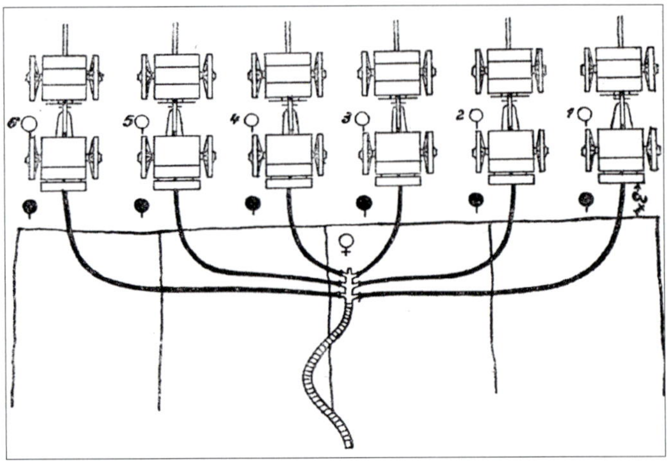

Above: The trolleys were filled according to a precise schedule.

Depending on the size of the balloon (ranging from 600-1000 m³), 120 to 200 gas cylinders were required which the division transported on six gas wagons. Gaskolonnen supplied Feldluftschiffer with gas replenishment; they, too, had six gas wagons. Gas replenishment from the homeland was assured by the Militär-Gas-Anstalt Brüssel (military gas works in Brussels), mobile gas works on train cars, as well as private companies. Throughout the course of the war and the increased demand for gas cylinders, Feldluftschiffer began utilizing trucks for transportation of people and goods. On trucks, gas bottles could be driven directly to the balloon and immediately hooked up to a filler hose. Ballonzüge 66 and 138 Gasmannschaften designed a wood construction which allowed a balloon to be filled directly from a truck, saving sweat and energy associated with physical labor. From July 1918, this construction was introduced and installed in all Gaskolonnen. With the enemy in sight, gas bottles were stored camouflaged near the ascent site and were connected to the filler hose via pipelines.

Transportation of gas bottles in mountainous terrain continued to pose a challenge. Therefore, Ballontruppen utilized Füllballone (filling balloons). These smaller balloons could easily be moved to the big ones where the gas was transferred. An enemy attack in the immediate vicinity of gas-filled balloons during gas transfer could easily have a catastrophic effect, which is why gas transfers usually took place in the early, dark hours of the morning.

The German Südarmee's FA (A) 242 was aware of the locations of Russian balloons' ascent sites due to aerial reconnaissance; on June 28, 1917, Offiziers-Stellvertreter Wilhelm Frickart departed for a balloon attack at 02:45:[13]

"I overflew enemy lines at 600 meters near Kuropatniki and glided toward the balloon which was anchored to the

Left: Each gas cylinder was connected to the collecting pipe via the red-painted copper pipes. The filling hose led from this to the balloon. (NARA)

Right: An FLA 8 balloon is filled with gas using a filling balloon.

ground near Kaplince. I attacked the balloon twice from 60 meters but was unable to set it ablaze. I then flew towards the south and attacked two balloons which were on the ground near Olesin, north of Helenkow. After one attack, the balloon caught on fire, simultaneously igniting the second balloon beside it, along with the accompanying four filler balloons.

"I returned to the front under continuing MG fire until I saw the two balloons on the ground near Telacze. I succeeded in setting both balloons ablaze from 300-400 meters in a single attack. I landed at the airfield at 03:50. The first attack was carried out at 03:00, the second one at 03:07, and the third one at 03:43. During the attack I was accompanied by an FA (A) 242 aircraft piloted by Lt. Hartwig who also observed the fires."

Lt. d.R. Wangemann, another member of FA (A) 242 also shot down a Russian balloon on September 21, 1917. The crew, consisting of Lt. d.R. Wangenmann and Uffz. Entenmann, wanted to destroy the enemy balloon on the ground prior to its ascent. They were familiar with the ascent site in a forested area because of aerial reconnaissance. Near Uwria, they descended to 1700 meters and approached the balloon in gliding flight to 150 meters. Wangenmann had a raised step added to the rear of the aircraft to facilitate vertical MG fire. During the first pass of the ascent site, Wangenmann hit the balloon with 50 rounds of phosphorus bullets. When Wangenmann and Entenmann overflew the site for the second time, they saw that smoke had formed on the balloon's outer gas bag. However, they encountered heavy MG fire which caused damage to the left wing strut, one wheel, and the elevator surface. On their return flight they were attacked by a Russian Nieuport; they were able to escape following a short aerial battle.

Telephones

Feldluftschiffer adapted the first, previously tested telephones from the Telegraphentruppe. The handsets initially used by observers were soon replaced by chest telephone receivers consisting of two headphones and a microphone attachable to the chest area. This hands-free telephone allowed observers the simultaneous use of binoculars, maps, and cameras. Fernsprechwagen 09 (telephone wagon) were utilized to connect the lines to the balloon headquarters and artillery. An insulated connecting cable between the balloon and the ground station simultaneously served to pass down message containers if a telephone system should become inoperative.

Cameras

L-30 / L-70 cameras with 30 and 70 cm focal distances, respectively, were developed for the Feldluftschiffer by Zeiss company in collaboration with Professor Miethe of the Technische Hochschule (technical university) in Charlottenburg from 1910. The size of the optical disc was 16x16cm. Major Hildebrandt and Hptm. Milczewski tested the cameras with Luftschiffer-Bataillon Nr. 1; upon completion of a few alterations, they were introduced to the Ballontruppe from 1911. For long-distance captures and panorama images of enemy territory, observers utilized the Zeiss L-120 camera with a focal length of 120 cm. During the war, Berlin cartographer, Karl Wulkow, developed a Grundrißbildner. Its use allowed for the conversion of oblique photographs into vertical aerial photographs and the creation of precisely scaled maps.

The Ballonzug 18 (w) panorama image was taken on August 17, 1917, on the ascent site near Buzy; it shows the southern Meuse valley. Twelve individual images were photographed at a 70-degree angle; pieced together they captured a 30km area of the French back country between Bouguemont and Fort Vaux.

The images were processed by Vermessungs-Abteilung 2 (surveying division 2) and once provided with location information, formed an aerial index map which was distributed among all units within that sector.

Above: The gas vans and a replacement basket of FLA 8 at Manancourt on the Somme.

Above: FLA 8 trucks in Merchem.

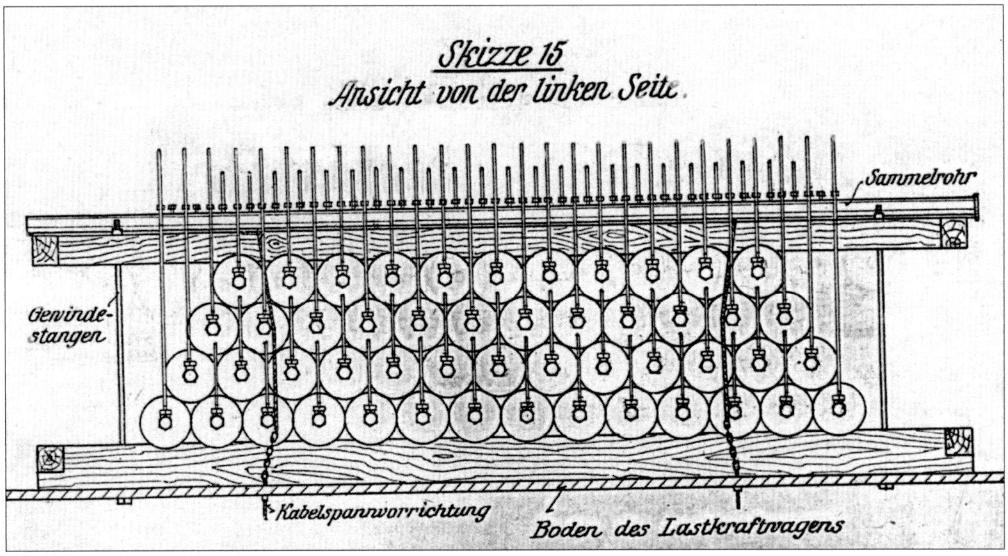

Right: The gas crews of balloon trains 66 and 138 constructed a wooden structure that allowed the balloon to be filled directly from the truck.

Right: Telephone car 09 was used to connect to the balloon control centers and the artillery.

Right: Telephone exchange 64 on the summer front.

Above: Early model of a balloon basket with a Zeiss L-70 camera.

Above: The airmen's ground telephone, on display in the Dübendorf Museum.

Albatros D.V *Mops*
***Offz. Stv.* Julius Buckler**
***Jasta* 17**
36 total victories, 7 balloons

Photos

Here are some examples of photos taken from balloons after they were printed and annotated.

Ballonbild
vom
Ballonzug 18(W)
aufgenommen am 17. 8. 1917.

Standpunkt 2882,5

Bearbeitet im Oktober 1917
bei der Vermessungs-Abteilung 2.

Right: Printed balloon flyer of the southern Meuse valley of Bz 18.

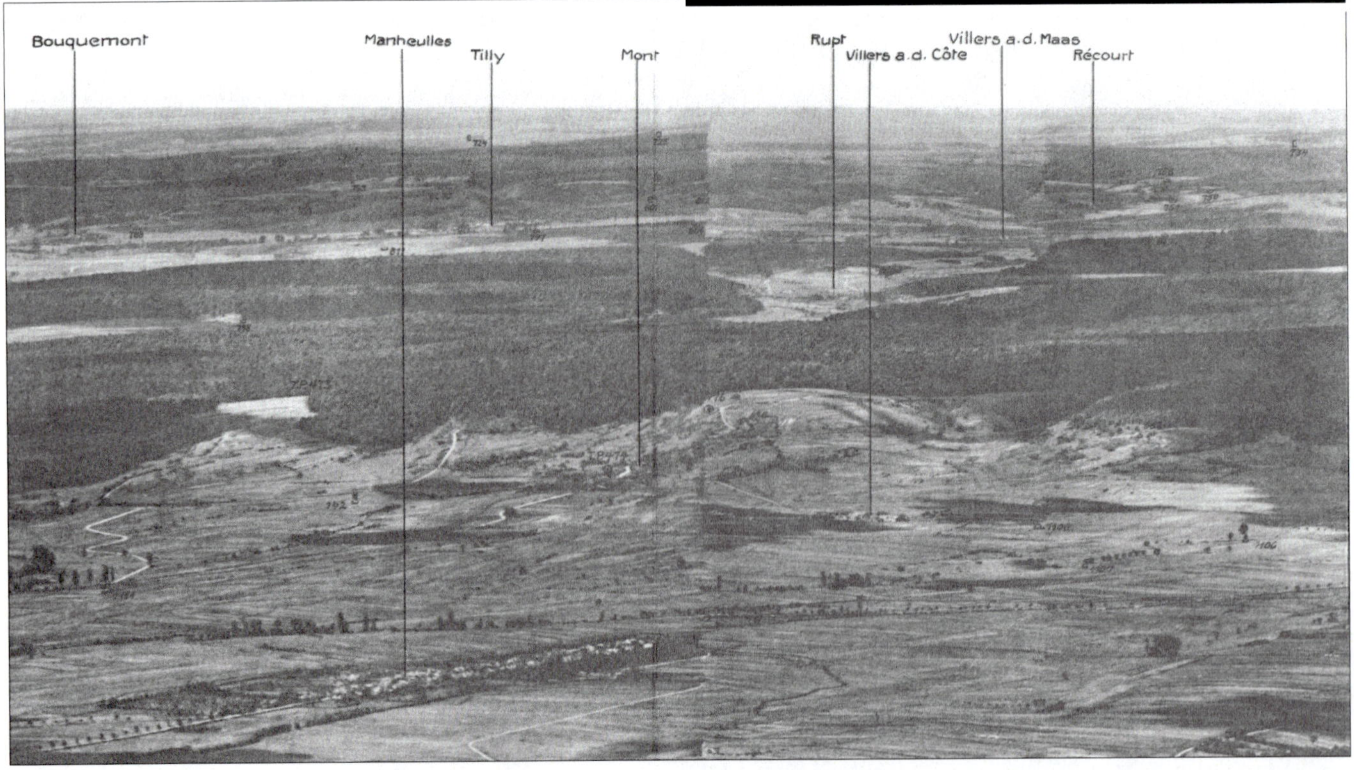

Above & Following Two Pages: Aerial photograph of balloon train 18 (Württemberg) on August 17, 1917 on the Verdun-Argonne front.

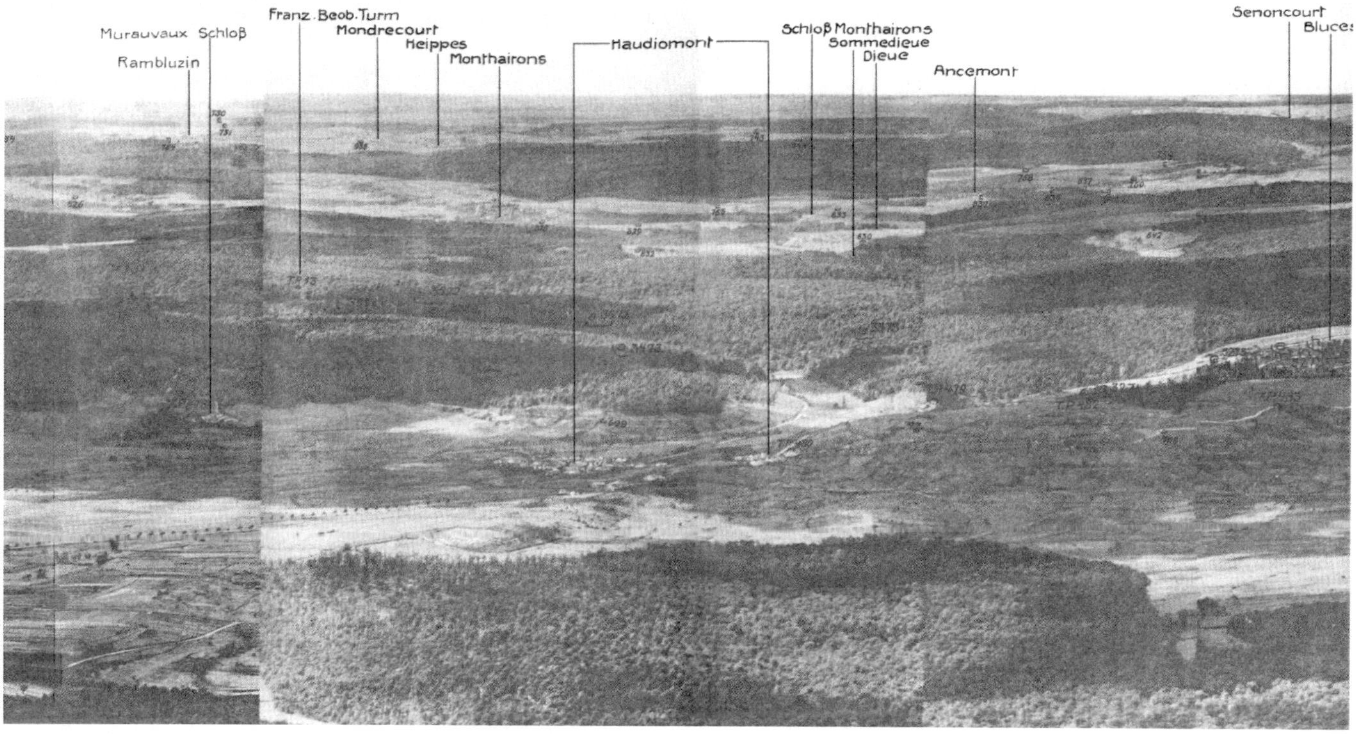
Murauvaux Schloß
Rambluzin
Franz. Beob.Turm
Mondrecourt
Heippes
Monthairons
Haudiomont
Schloß Monthairons
Sommedieue
Dieue
Ancemont
Senoncourt
Bluces

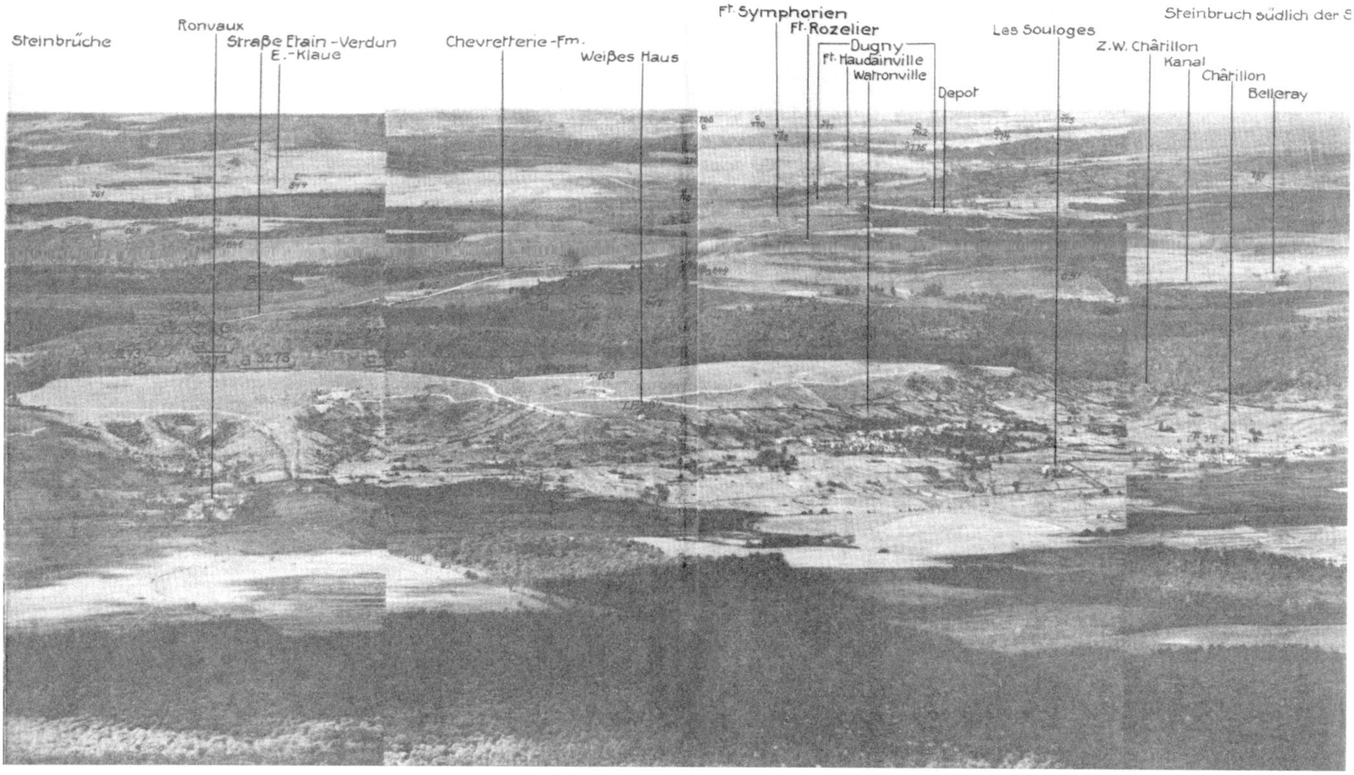
Steinbrüche
Ronvaux
Straße Etain-Verdun
E.-Klaue
Chevretterie-Fm.
Weißes Haus
Ft. Symphorien
Ft. Rozelier
Dugny
Ft. Haudainville
Watronville
Depot
Les Souloges
Steinbruch südlich der S
Z.W. Châtillon
Kanal
Châtillon
Belleray

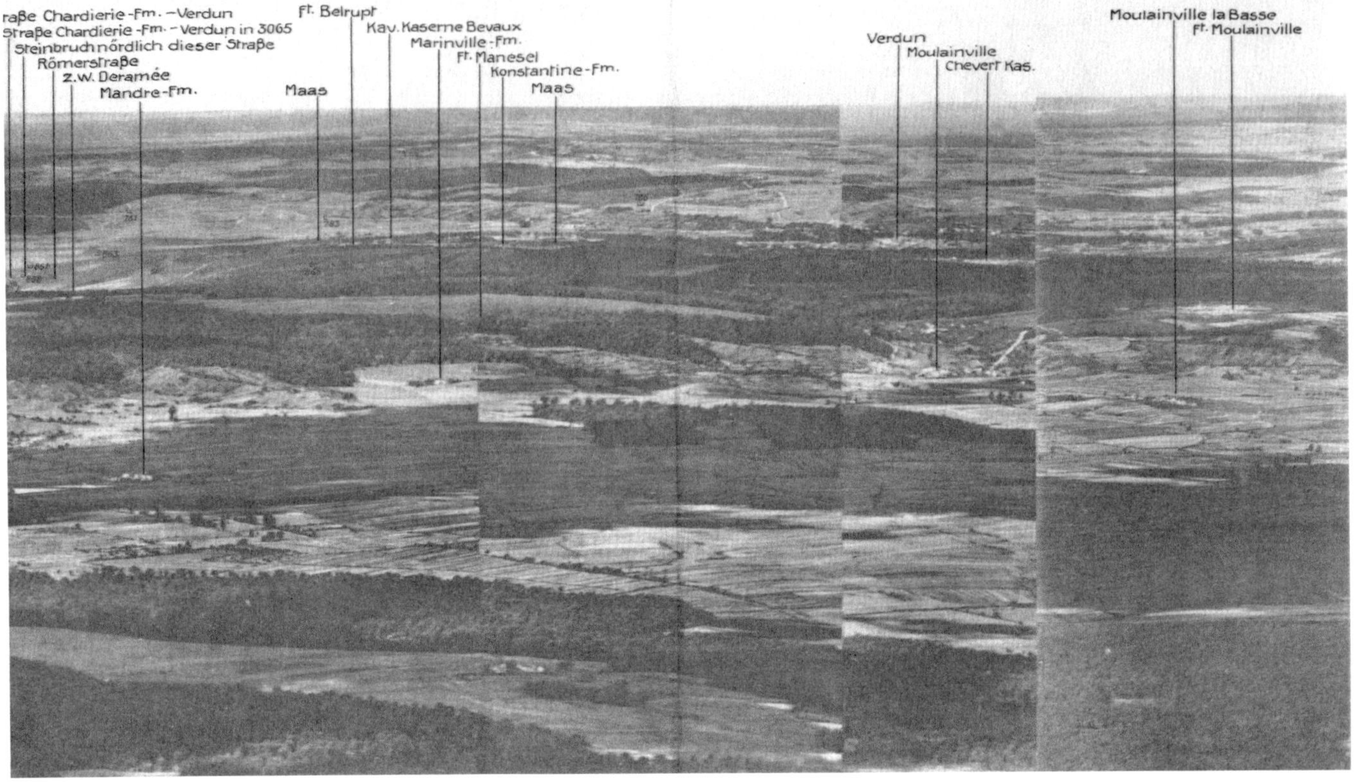

raße Chardierie-Fm. – Verdun
Straße Chardierie-Fm. – Verdun in 3065
Steinbruch nördlich dieser Straße
Römerstraße
Z.W. Deramée
Mandre-Fm.
Maas
Ft. Belrupt
Kav. Kaserne Bevaux
Marinville-Fm.
Ft. Manesel
Konstantine-Fm.
Maas
Verdun
Moulainville
Chevert Kas.
Moulainville la Basse
Ft. Moulainville

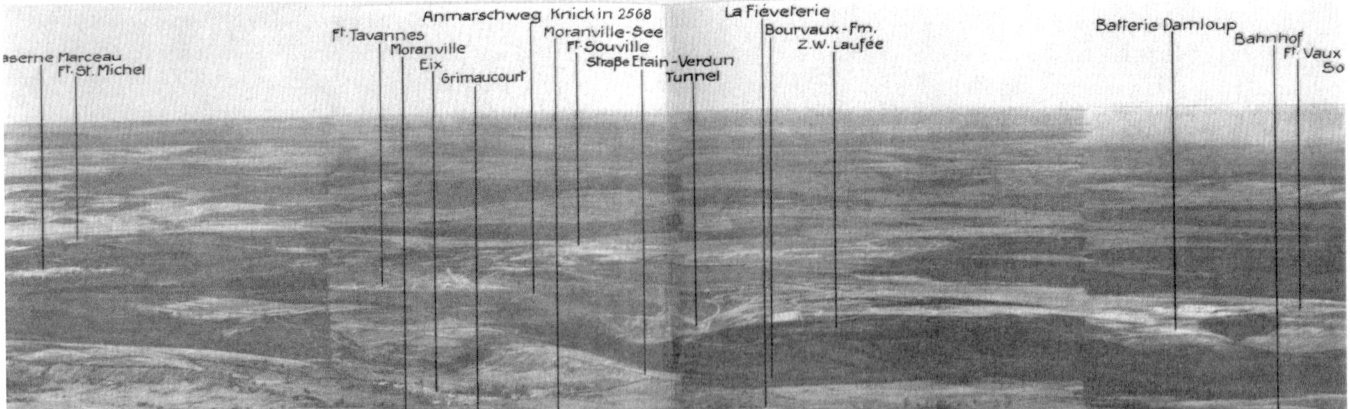

aserne Marceau
Ft. St. Michel
Ft. Tavannes
Moranville
Eix
Grimaucourt
Anmarschweg
Knick in 2568
Moranville-See
Ft. Souville
Straße Etain-Verdun
Tunnel
La Fiéveterie
Bourvaux-Fm.
Z.W. Laufée
Batterie Damloup
Bahnhof
Ft. Vaux
So

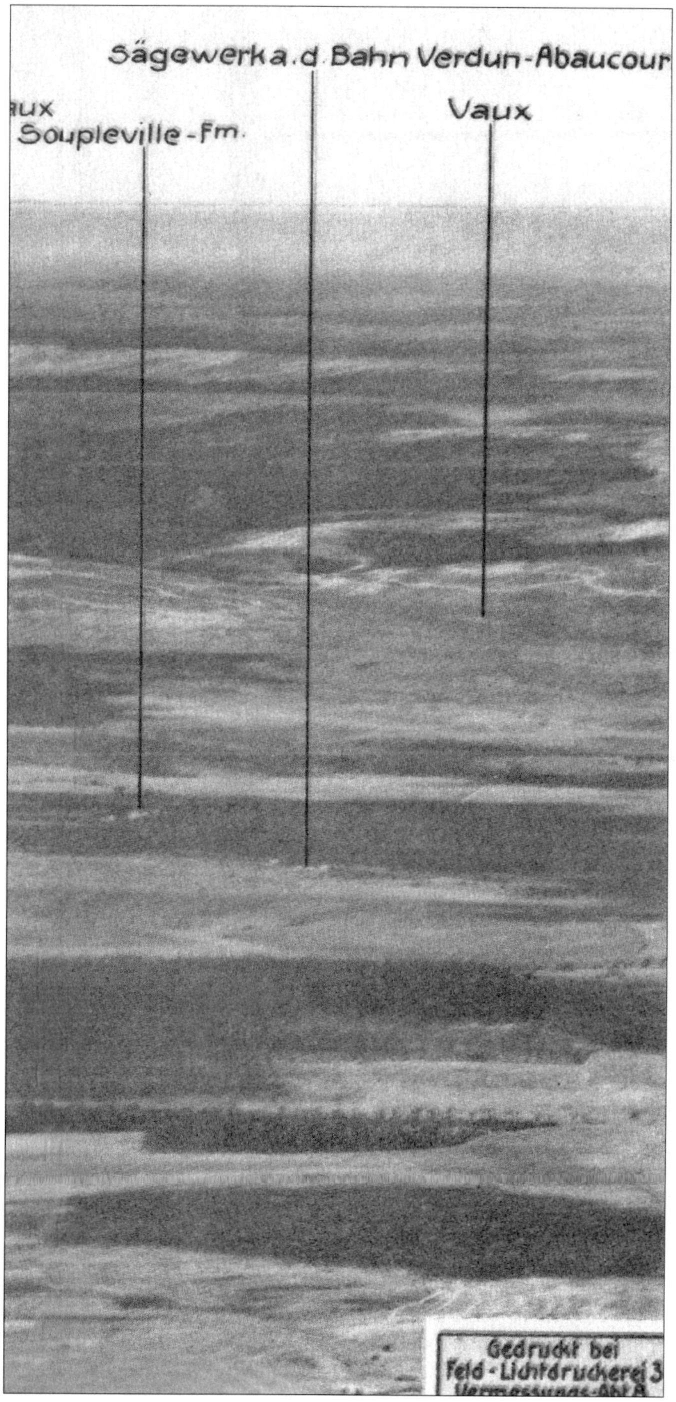

Above: Aerial photograph of balloon train 18 (Württemberg) on August 17, 1917 on the Verdun-Argonne front.

Above: Balloon observer with the Zeiss 120 cm camera, no easy task in the narrow basket. The observer wears the simple parachute harness around his chest. (NARA).

Above: At Orfenil of Bz 50, the balloon is ready to rise. The large camera protrudes far beyond the basket.

Above: The Bavarian FLA 8 was part of the Loos Group in Flanders in April and May 1917. Lt. Wagenscheid photographed the neighboring balloon here (arrow).

Right: The Swiss balloon troop also used French balloons and German Zeiss cameras.

Above: The picture plates were also processed by professional photographers in the picture offices and copies were also made.

Above: Test ascent with the 120 cm camera. There is no parachute attached to either the basket or the lines. (NARA)

Parachutes

Above: Two parachutes are attached to the cross strut in bags above the gondola. The observer only clipped into the parachute line before the jump.

Above Left: The Paulus parachute saved the lives of many observers. In this color picture, however, the second observer jumps out too late. The balloon is already on fire and his parachute was therefore safely caught by the fire.

Observer jumping out of the basket. (NARA)

The French André-Jaques Garnerin is said to be the inventor of the parachute. On October 22, 1897, he jumped out of his balloon with a hand-sewn parachute from an altitude of approximately 1000 meters and landed safely on the ground. In 1910, the Austrian Reichelt tested a parachute suit which was too small. On February 6, 1912, Reichelt jumped from the Eiffel tower with his parachute and sustained fatal injuries.

On March 1 and 10, 1912, the American Berry was more successful when he jumped out of a biplane in St. Louis. The parachute was stored in a bag beneath the wings and landed smoothly. On August 19, 1913, Pégoud successfully jumped out of a Bleriot monoplane.

Käthe Paulus parachutes became the Ballontruppe's standard parachutes. Paulus, a dressmaker by trade, incidentally, witnessed a parachuter jump out of a free

Above: Captain Ritter Adolf von Tutschek in the spring of 1918.

Above: Hptm. Tutschek lands with his Fokker Dr. I 404/17 in Toulis.

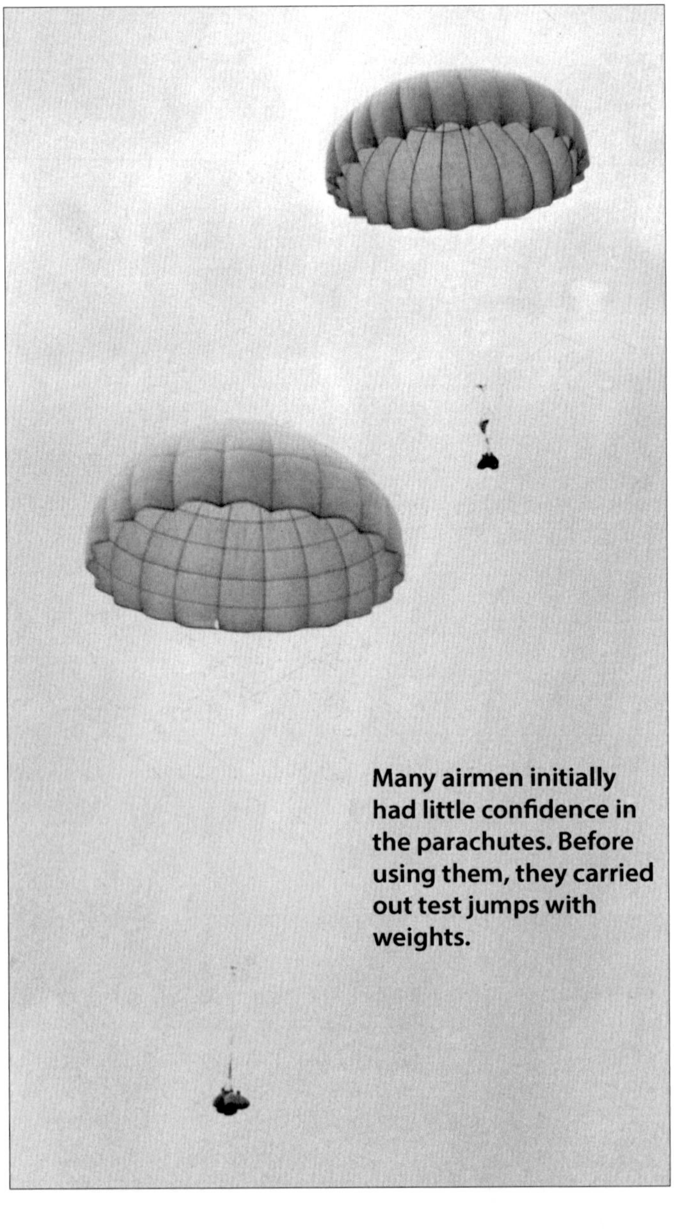

Many airmen initially had little confidence in the parachutes. Before using them, they carried out test jumps with weights.

balloon in 1892. Impressed by this experience, the young woman founded a company for balloon outer bags and parachutes with her future husband, Hermann Lattemann. Between 1893 and 1914, Paulus herself jumped out of balloons with parachutes more than 147 times. After Hermann Lattemann was killed in a balloon accident on June 17, 1894, she continued to lead the company alone.

At the 1909 international Luftschifffahrt-Ausstellung (airship travel exhibition) in Frankfurt, she had her own stand for the first time. From 1912, Käthe Paulus manufactured balloon gas bags in Berlin-Reinickendorf; from 1915, she also began manufacturing parachutes. The Paketfallschirm (package parachute) was her invention which

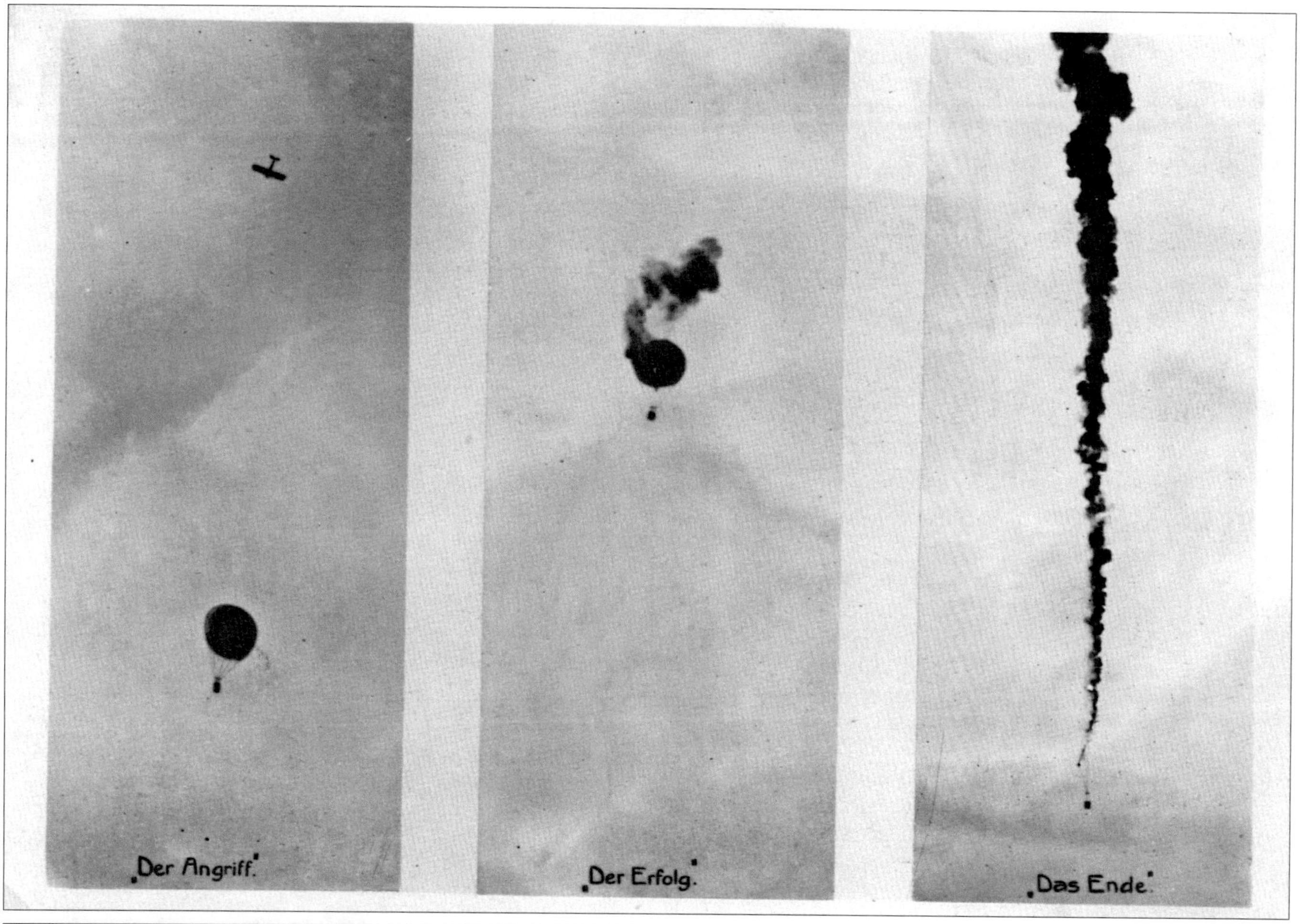

Above: The effect of phosphorus projectiles was tested by Idflieg on obsolete balloons.

Jump of a British observer with a Spencer-Guardian-Angel parachute.

she sewed in her apartment supported by seamstresses who worked from their homes.

By the spring of 1915, Paulus offered her parachutes to the Inspektion des Militär- Luft- und Kraftfahrwesen who turned down her proposal. It was not until Ballontruppe losses rose that the agency reflected on Paulus' offer and hastily placed a parachute order. To produce a large quantity, Paulus was provided with a large warehouse in Berlin Reinickendorf where approximately 7000 parachutes were manufactured by the end of the war.

The parachutes were placed into small bags and attached either to the basket or to the balloon baskets' carrier ropes. Initially, observers wore a simple belt which was later supplemented by shoulder straps and a pair of parachute trousers. It was not a rare occurrence for observers to jump out of a balloon basket multiple times a day. Of course, every jump put stress on the observers' nervous system; soldiers in the trenches were facing the same problem before every assault. Ultimately, the introduction of the parachute saved the lives of many observers, Allied Forces and Germans alike. While there was still a risk of injuries during hard landings or being dragged on the ground with a partially opened parachute, but they knew this was a far better alternative to

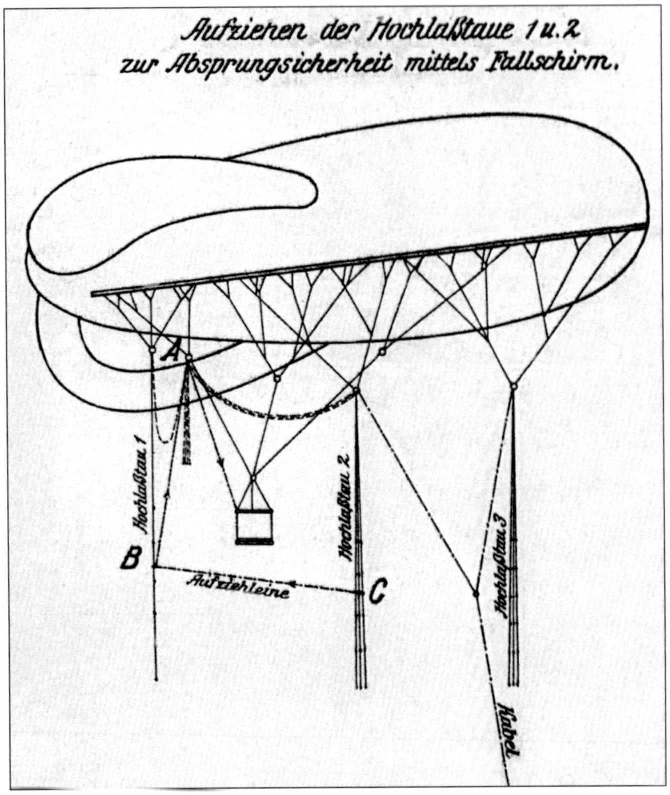

Above: The balloon's lines were not allowed to obstruct the observer during the jump, so they had to be attached at a safe distance from the basket.

facing certain death by jumping out of the basket without a parachute or being burned to death.

Feldluftschiffer-Abteilung 39 Vizefeldwebel Hausmann wrote a report about his jump on August 6, 1916:[14]

"When I was ordered to prepare for the jump, I had not yet noticed the enemy aircraft. I calmly clasped the carabiner hooks into the belt rings and awaited further instructions. At the same time, I heard the popping of bullets overhead. I removed all the items that could pose a hindrance during the jump, sat on the edge of the basket, holding the phone apparatus in my hand. When orders were given to jump, I briefly hesitated but jumped out quickly after it became clear that the balloon had been hit.

An observer fires signals to the infantry with the signal pistol. The parachute is attached to the basket.

"I don't have any memory leading up to the moment the parachute opened. I suddenly became aware of the parachute above me and felt completely safe. The parachute was lowering me down slowly, seemingly endlessly. I was afraid the exploding anti-aircraft grenades and MG bullets shot from aircraft would hit me.

"When I was about 150 meters above the ground, I noticed that the parachute was falling considerably faster. When I looked up, I saw a small tear in the fabric. Convinced that I was about to break my legs, I pulled them up to my torso; as I hit the ground, I rolled onto my side. Initially, I thought I fractured a bone as I felt a lot of pain in my right leg. After multiple attempts, I got up and loosened the strap. By that time, crews and a medic were approaching."

In an emergency, even the observers' valuable cameras could be dropped out of the balloon basket with a parachute. Hauptmann Ritter v. Tutschek described this occurrence in a letter dated March 1, 1918:[15]

"This morning at 08:45 I attacked a French balloon near Terny from 3000 meters in my Fokker Dr. I 404 (/17). After firing 10 rounds Ph. L (phosphorus and incendiary ammunition), the balloon, which was at 1300 meters, caught on fire. A large, black box (camera) with a red parachute attached was tossed overboard, followed by two observers wearing white parachutes."

While parachutes saved lives, some observers were still hesitant to put on parachute belts in 1916. They felt constricted by the belt while carrying out their tasks in the tight balloon basket. On June 26, 1916, Lt. d.R. Böhme was tasked with surveying the front area south of the river Somme using a 120 cm camera. Böhme once again planned on ascending without a parachute but this time, his battalion commander insisted on him putting on the parachute belt; the parachute bag was attached to the balloon basket. One hour after the ascent, ground crews reported that the adjacent balloons had caught on fire. A motorized winch was used to begin the pulling down of the balloons from a height of 1200 meters while anti-aircraft gunners attempted to drive away the attacking French aircraft.

After a few seconds, flames began to surface on Lt. Böhme's balloon; he jumped out of the basket. Following 60 meters of freefall, Böhme's parachute opened. However, he had not adjusted the belt to his waist size; it was much too wide for his body, and he began to slide out of the harness. Böhme struggled to hold on until safely touching down on a nearby road. In the early stages of parachute utilization, the British, too, recorded several parachute accidents. Basil

Hallam of the 23rd Section, commanded by Captain David Bauer, fell to his death when his parachute failed to open. After this accident, observers personally folded their own parachutes, placing sheets of paper in between individual fabric layers.[16]

A successful jump out of a balloon basket did not necessarily mean a safe outcome of a bad situation. On June 3, 1917, the Bavarian FLA 4 was stationed near Moussey, with its subordinate Ballonzügen 10, 11, and 12. Balloon observer Vzfw. Janus was about to begin zeroing in Artillerie-Batterie 864 on enemy targets:[17]

"When two enemy aircraft approached the balloon at an altitude of 1000 meters, they faked an aerial battle and suddenly attacked the balloon using MG flame projectiles. When they aircraft were only about 100 meters away from the balloon, the MG fire increased. I did not hear any anti-aircraft gun activity, so I cut the cord to the chest telephone and jumped out of the basket from about 700 meters up.

"In my evaluation, the parachute opened after approximately 10 meters and slowly glided towards the ground. While falling, I had time to loosen the disturbing chest telephone and, after twisting my body somewhat, I was able to observe the burning balloon.

One of the aircraft pursued me until about 400 meters; he shot at me with his MG from no further than 100 meters. If I had had a weapon, I would have returned fire."

Initially, the ground crew Offizier, Oblt. Kaufmann did not notice the aircraft, nor did he hear any engine noises. As a nearby anti-aircraft gunners had opened fire, Kaufmann gave orders to bring down the balloon. When the balloon was at 900 meters, anti-aircraft gunners reported that aircraft, likely German, were approaching. The winch was stopped; shortly after, Oblt. Kaufmann spotted the two Escadrille N77[18] at 1600 meters, aiming directly for the balloon. He immediately had the winch engaged at full power; the cable, pulled by six horses, was placed over a castor to speed up the process of bringing down the balloon. Meanwhile, Vzfw. Janus jumped out of with the parachute. The two Nieuport 17 were shooting at the balloon. Flying at low level, SLt Maurice Boyau set the balloon ablaze while the second pilot shot at the observer floating down in his parachute.[19]

Despite MG fire and the fire of Maschinen-Flakzug 28, balloon No. B 11 partially burned. 14 German observers in total died by enemy aircraft fire while approaching the ground with their parachutes.

Two days later, SLt Maurice Boyau returned to visit FLA 4 with a comrade. From Manoncourt-en-Vermois airfield near Nancy, two Nieuport 17 approached the German balloons at high level. At 12:25, while Abteilungsführer Hptm. Rötscher discussed the site of the future balloon hangar with Bz 12 Lt. Seitz, both men noticed the bright glistening of the French aircraft, one of which was firing shots at the B 12 balloon. From an altitude of 200 meters, observer Vzfw. Meyer jumped out of the balloon with his parachute and landed safely. The balloon was engulfed in flames shortly after. Vzfw. Meyer had attempts at hitting the Nieuport with his Mauser pistol were futile. The Abteilung's two 3.7cm anti-aircraft machine guns opened heavy fire on the Nieuport which was flying directly towards the front lines. Meanwhile, the other Nieuport initiated a fake attack on the Bz 10 balloon, drawing the MG and anti-aircraft gun fire away from his comrade. According to observations from the Bz 11 balloon, both Nieuport were able to escape across the front lines.

Abteilungsführer Hptm. Rötscher criticized the weakness of FLA 4's anti-aircraft gun protection, fearing the French would want to prevent balloon observations at all costs. According to his opinion, balloon specialists would be utilized for this purpose, such as Escadrille 77 SLt Maurice Boyau who had shot down 21 balloons.

In the summer of 1918, Inspektion der Luftschiffertruppen, Ballonmeisterei division, tested the Schmittner type Korbfallschirm (basket parachute). With this type of parachute, observers no longer had to jump out of the basket with the parachute but instead were able to float down to the ground in the balloon basket. On August 6, 1918, Gefreiter Paul Schneider tested the basket parachute in Berlin-Reinickendorf. The balloon basket, observer, and camera weighed 172 kg, the parachute itself 31 kg. At 400 meters, Schneider released the basket parachute by pulling the red handle on the balloon's suspension. Freed from the weight of the basket, the balloon shot into the air, triggering the basket parachute to open after two seconds. Schneider reported on the event:[20]

"I felt like I was slowly floating downward in an elevator. I did not have to hold on to anything while descending, waved my hand at the people below. I particularly enjoyed not being strapped in or feeling the tight parachute belt around my waist. [...] I would much rather float down to the ground ten times in a basket parachute than one time with a regular parachute."

The Offiziere der Inspektion der Luftschiffertruppe (Idluft) watched the practical testing of the basket parachute from the ground. Among them was Idluft Chef, Major Gundel. Gefreiter Schneider composed a detailed report about the basket parachute, to which Major Gundel added his own comments. From August 1918, all Ballonzüge were equipped with basket parachutes.

Flak (Anti-Aircraft Guns)

Above: An MG 08 on rotating mount near Loupy in Lorraine. (NARA)

Above: The rotating mounts also appear to have been manufactured by the industry, as they were frequently used. (NARA)

Above: A 3.7 cm Maschinenflak on Daimler truck of Bz 50.

Above & Facing Page, Above: The protectors of the balloons, a 7.7 cm Flak L/27 on a Krupp truck.

Until 1915, only carbine rifles were utilized for anti-aircraft artillery. Those rifles only had a moral effect, however. Hits on an enemy aircraft from the ground or from a balloon were not achieved with carbine rifles. It was not until Ballontruppen were issued MG 08 or Beute MG that they were successful in striking the enemy from the ground or from a balloon. Furthermore, 3.7 cm anti-aircraft machine guns and mobile anti-aircraft guns on motor vehicles measuring 7.7cm in caliber (Kraftwagen-Flak L/27 c/1914).[21]

Until March 1915, Krupp Maschinenflak served as the onboard gun of choice for torpedo ships and smaller naval vessels. Following the re-arming with stronger calibers, the Krupp company in Essen converted a total of 216 cannons into guns serving only the purpose of anti-aircraft artillery. Using Granade-catridges with flare kits produced a visible fire trail which also had a psychological effect. These guns were capable of firing 250 rounds per minute with a range of 6000 meters. The heavy, 8.8 cm and 10.5 cm caliber, mobile anti-aircraft artillery (Kraftwagen-Flak or sK-Flak) were deployed to the front from late 1916.[22] With a range of 10,000 meters the guns placed a barrier in front of the balloons which forced most fighter aircraft to change course. The Bz 94 (FLA 7) war diary contains a list of 17 unsuccessful air attacks which had been fought off by heavy mobile anti-aircraft artillery 109 (commanded by Hptm. Brehm). On May 29, 1918, artillerists discovered three Spad approaching a balloon; the aircraft discontinued their attack due to the barrage fire. On June 1, the Bz 94 marched through Amentieres in the Ballon hoch configuration. Bringing in balloons in a city was impossible; furthermore, Spad were attacking. SK-Flak 109 and Maschinen-Flakzug 178 immediately set up a barrage fire in front of the balloon, allowing it to march onward to Lattily unharmed.

On April 29, 1917, No 40 Squadron Lt JAG Brewis attacked a Bavarian FLA 1 balloon. Balloon Nr. B.A.2R. 2450 was at a height of 1000 meters; at 08:20, after a

Above: French 7.5 cm anti-aircraft gun on Delahaye truck.

few rounds fired by Lt. Brewis, the balloon was on fire. Meanwhile, anti-aircraft artillery fired their guns and MG at Nieuport 17 (A6739). Lt. JAG Brewis was fatally hit by a 3.7cm anti-aircraft automatic weapon and plummeted to the ground.

On October 18, 1917, Jasta 29 Oblt. Otto Schmidt found himself under heavy attack by the British balloon defense in the 4th army's sector in Flanders. Schmidt was returning from a visit at Jasta 11's base at the Aatrijke airport with his Staffel comrade, Lt. Fritz Kieckhafer. They planned to attack enemy balloons along the front:[23]

"On October 18, 1917, at approximately 13:30, we overflew the front towards the enemy near Ypern at an altitude of 4000 meters. We flew until we were about 12 kilometers behind the lines and saw the balloons below us at an angle. I gave the agreed-on signal, and we began our attack by entering a nosedive. Kieckhafer was lucky; the balloon beneath him was at a high altitude, allowing him to catch up to it at approximately 500 meters despite ground crews' attempts to bring it in. He set the balloon on fire and returned as previously discussed.

"I hesitated too long as I wanted to observe and secure my comrade's attack. When I finally approached my balloon, it had already been lowered considerably, forcing me to descend to 200 meters. I was hit in the lower leg before I was able to open fire on the balloon. As would later be discovered, an MG bullet fired by ground defense pierced my shin and fibula below the knee. Turning the aircraft became a challenge as I was unable to move the rudder pedal. As I was heading east, back to our lines, I pulled the aircraft from its nosedive and flew home as fast as I could.

"However, the enemy's balloon protection became aware of my attack and cut off my return flight. Three aircraft approached me and began their attack. I was no longer able to engage in a dogfight. I am certain I would have been forced to a landing in enemy territory due to excessive blood loss. I had no choice but to fly straight at maximum power.

"I heard crashes and bangs as enemy fire hit my aircraft. A grazing shot hit my arm."

Finally, German aircraft came to Schmidt's aid and drove the British away. Badly wounded, he managed to land at the Jasta 29 field. Oblt. Schmidt did not return to his Staffel until July 3, 1918. Through summer of 1918, Jasta 29 aircraft colors included a yellow nose and green fuselage. A "Chinese dragon" adorned both sides of the aircraft's fuselage.[24]

Even the strongest anti-aircraft artillery was not able to protect the balloons when personnel were not ready for battle. On February 2, 1918, Bz 6 (FLA 5) made this unfortunate experience near Villers-Outreaux. In the morning, Lt. Gmelin had the balloon brought from the hangar to the ascent site on the edge of the Mortho forest and ordered for it to be anchored. At 15:05, an enemy aircraft attacked the balloon from a height of 10 meters. The balloon was on fire after the fourth attack. Attempts by ground crews to drive the aircraft away using carbine rifles remained unsuccessful. Bz 6's MG and anti-aircraft automatic weapons were in their quarters and did not partake in the defense.[25] Following this incident, Kogenluft ordered for continuous occupation of MG and anti-aircraft artillery during daylight hours.

Right: German 37 mm Flak gun and crew.

Above: Oblt. Otto Schmidt of Jasta 29 came under heavy fire from the British balloon defense in Flanders with the 4th Army on October 18, 1917.

Above: Lt. Siempelkamp in front of his yellow and green Pfalz D IIIa at Jagdstaffel 29 with his personal marking, a Chinese paper dragoon.

Lt. Eugen Siempelkamp, Pfalz D IIIa,
Jagdstaffel 29,
Late Spring–early Summer 1918

FLA 37 Photo Album

Above: Leutnant Ruff of the FLA 37, on the epaulette is an "L" for "Luftschiffer."

A few years ago, I was able to purchase a Ballon-Abteilung photoalbum in France. Unfortunately, no captions were placed next to the photographs. There are dates on the back of the pictures; however, super glue was used to attach the pictures to the pages which does not allow the removal of the images without destroying them. There are multiple pictures in which Offiziere wear FLA 37 badges which means the photos must be of Feldluftschiffer-Abteilung 37. The only town referenced is Les Salieres. The Caudron R 11 with registration No. C 1636 must have landed in the vicinity of FLA 37. Maybe one of the readers has some knowledge of FLA 37 and can provide information on the subject.

Above: The participants in the airship pilot course, probably in the winter of 1915.

Above: The commander of FLA 37.

Above: FLA 37 being moved by train.

Above: Retrieving the balloon of FLA 37.

Above: In May 1916, FLA 37 was at Les Salieres.

Above: The department's MG 08 was mounted on a cart.

Above: The gas from the envelope is emptied.

Above: The French population was naturally very interested in the work of the balloon troops.

Above: Balloon soldiers with the children of their host family.

Above: Members of FLA 37 in front of the accommodation.

Above: Ascent at FLA 37; the observers are not yet wearing a parachute harness.

Right: Burning balloon of FLA 37.

Above: Caudron R 11 No. C1636 is looked at and photographed by German soldiers.

Above: Group photo of the airmen in Flanders. The officer in the front row is wearing the airman's commemorative badge. Members of the navy can also be seen in the top row.

Left: An American victory in the Maas-Argonne Battle.

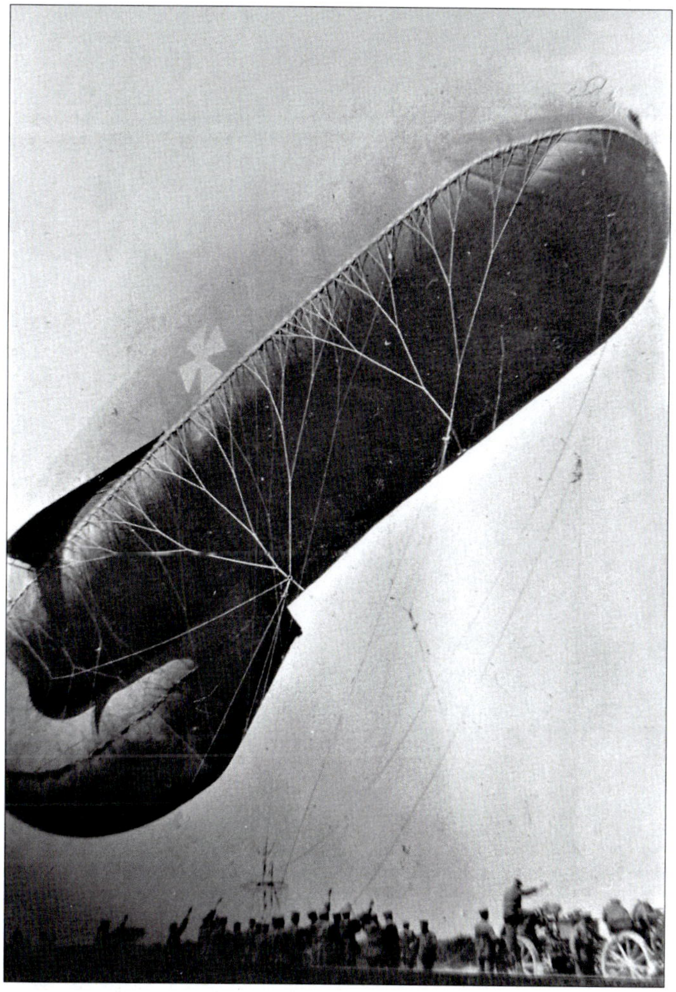

Above: A photo from the Jasta 15 photo album shows a German balloon ready to take of.

Above: British balloon on transport.

Above: German balloon at the Chemin des Dames.

Endnotes

1. Peter, Ernst: *Die k.u.k. Luftschiffer- und Fliegertruppe Österreich-Ungarns*, Stuttgart 1981.
2. Neumann, Georg Paul: *Die deutschen Luftstreitkräfte im Weltkriege*, Berlin 1920.
3. American Réne Gregory from Philadelphia was the second oldest soldier in the German army. He fell near Neufchátel at the age of 70 (Infanterie-Regiment 106).
4. Méchin, David: *The WWI French Aces Encyclopedia*, Vol. 2, Reno 2021.
5. Neumann, Georg Paul: *Die deutschen Luftstreitkräfte im Weltkriege*, Berlin 1920, p. 580.
6. *Jahrbuch der Luftwaffe*, Berlin 1936.
7. Kirchner, Wilhelm: *Feldballon und Luftsperren*, Berlin 1939.
8. *Unsere Luftwaffe im Weltkrieg*, Leipzig o. J.: Lt. d. L. Cohn, Bei den Feldluftschiffern.
9. Bavarian National Archives, Department IV War Archives: File MKR 1363.
10. Ballonhüllengesellschaft mbH, Berlin Tempelhof was part of Luftschiffbau-Zeppelin GmbH.
11. ARA - *A. Riedinger Ballonfabrik AG 1897-1922*, München 1922.
12. Bavarian National Archives, Department IV War Archives: File ILuft 40
13. Air Force Bulletin Number 20, July 1917.
14. Wir Kämpfer im Weltkrieg: Bericht des Vizefeldwebels Hausmann.
15. Tutschek, Adolf Ritter von: *In Trichtern und Wolken*, Braunschweig 1934.
16. Captain GD Machin: Kite Balloon Recollections, *Cross & Cockade* GB Vol. 4 Farnborough 1973
17. Bavarian National Archives, Department IV War Archives: File ILuft 40.
18. Escadrille N 77 was stationed at Manoncourt-en-vermois.
19. Méchin, David: *The WWI French Aces Encyclopedia*, Vol. 1, Reno 2021.
20. Bavarian National Archives, Department IV War Archives: File ILuft 41.
21. Spielberger, Walter: *Deutsche Flakpanzer bis 1945*, München 1982.
22. Grimme, General: *Ehrenblätter der deutschen Flakwaffe*, Berlin 1940.
23. Schmidt, Otto in Brzenk, Hans: *Wir jagten den Feind,* Berlin 1939.
24. Schmäling, Bruno: *Jasta Colors* volume 1, Reno 2021.
25. Bavarian National Archives, Department IV War Archives: File ILuft 41.

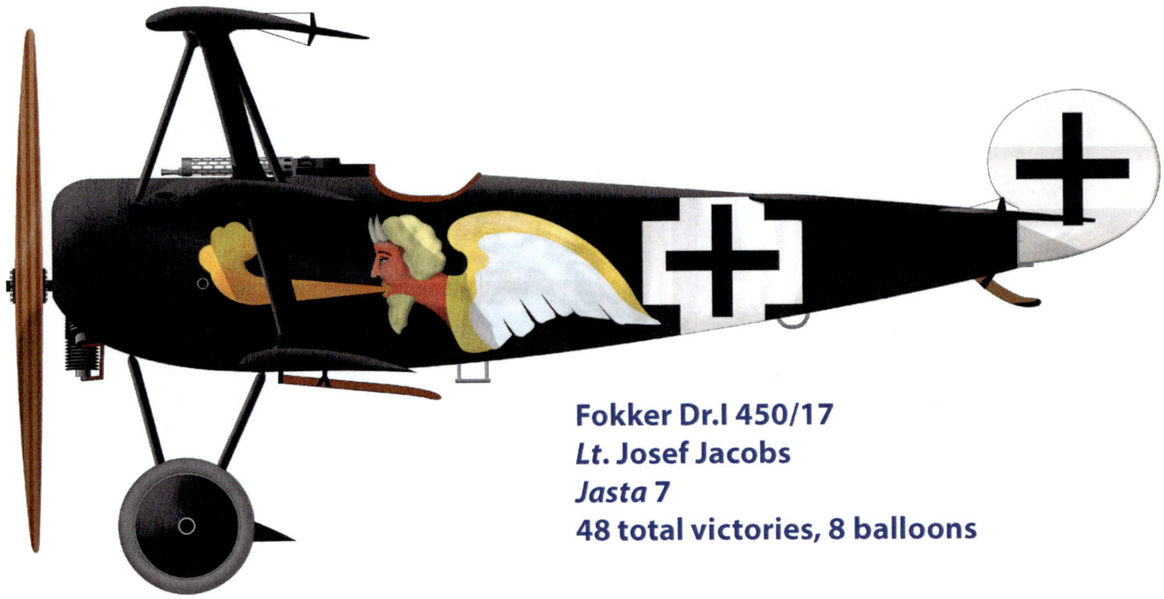

Fokker Dr.I 450/17
***Lt.* Josef Jacobs**
***Jasta* 7**
48 total victories, 8 balloons

Pfalz D.IIIa
***Lt.* Erich Lowenhardt**
***Jasta* 10**
54 total victories, 9 balloons

Fokker D.VII(Alb) 5278/18 *Hertha*
***Lt.* Friedrich Noltenius**
***Jasta* 27**
21 total victories, 8 balloons